**WITHDRAWN
UTSA Libraries**

An Environmental History of Postcolonial North India

POSTCOLONIAL STUDIES

Maria C. Zamora
General Editor

Vol. 2

PETER LANG
New York • Washington, D.C./Baltimore • Bern
Frankfurt am Main • Berlin • Brussels • Vienna • Oxford

Eric A. Strahorn

An Environmental History of Postcolonial North India

The Himalayan Tarai in Uttar Pradesh and Uttaranchal

PETER LANG
New York • Washington, D.C./Baltimore • Bern
Frankfurt am Main • Berlin • Brussels • Vienna • Oxford

Library of Congress Cataloging-in-Publication Data

Strahorn, Eric A.
An environmental history of postcolonial North India: the Himalayan
Tarai in Uttar Pradesh and Uttaranchal / Eric A. Strahorn.
p. cm. — (Postcolonial studies; v. 2)
Includes bibliographical references and index.
1. Tarai (India and Nepal)—Environmental conditions. 2. Tarai (India
and Nepal)—History. 3. Land use—Social aspects—Tarai (India and Nepal)—History.
4. Land use—Political aspects—Tarai (India and Nepal)—History. 5. Land
use—Economic aspects—Tarai (India and Nepal)—History. 6. Uttar Pradesh
(India)—Environmental conditions. 7. Uttaranchal (India)—Environmental
conditions. 8. Landscape—India—History. 9. Postcolonialism—Environmental
aspects—India. 10. India—Politics and government—1947– I. Title.
GE160.I4S77   333.70954'2—dc22   2009017068
ISBN 978-1-4331-0580-7
ISSN 1942-6100

Bibliographic information published by **Die Deutsche Bibliothek**.
Die Deutsche Bibliothek lists this publication in the "Deutsche
Nationalbibliografie"; detailed bibliographic data is available
on the Internet at http://dnb.ddb.de/.

The paper in this book meets the guidelines for permanence and durability
of the Committee on Production Guidelines for Book Longevity
of the Council of Library Resources.

© 2009 Peter Lang Publishing, Inc., New York
29 Broadway, 18th floor, New York, NY 10006
www.peterlang.com

All rights reserved.
Reprint or reproduction, even partially, in all forms such as microfilm,
xerography, microfiche, microcard, and offset strictly prohibited.

Printed in Germany

To My Family

Contents

Foreword ix

List of Abbreviations xi

Introduction 1

The Ecology of the Tarai 7

The Production of Exoticism 19

G.B. Pant and The U.P. Government 43

Realizing the Ideal 59

Outside the Official Paradigm 85

The Changing Tarai Landscape 101

Conclusion 135

Bibliography 141

Index 163

Foreword

I have completed this project only with the advice, criticism, and support of many people. Most important are my adviser, Paul Greenough, and the members of my dissertation committee: Jim Giblin, Jeff Cox, Jael Silliman, Philip Lutgendorf, and Stephen Vlastos. I would also like to thank Mridula Mukherjee, my adviser at Jawaharlal Nehru University, and Amiya Kesavan, my case officer at the United States Educational Foundation in India. In addition, I must thank Arindam Das and O.P. Bhardwaj of USEFI, Uma Narain and the late K.C. Sharma of Lucknow, Bob Alter of Mussoorie, Romesh Chandra of the Forest Research Institute (retd.), Raghubir Chand and Shekar Pathak of Kumaun University, G.S. Joshi, Assistant Commissioner of Nainital District, S. Moitra, Chief Librarian, Indian Council of Agricultural Research Library, Ramachandra Guha, Satyesh Chakraborty, Pramod Parajuli, Richard Tucker, Michael Mann, Clea Finkle, William J. Alspaugh, James Nye, Aijazuddin Ahmad, Irvin D.S. Winsboro, Nicola Foote, John Cox, Michael Epple, Michael Cole, Erik Carlson, Megan McShane, Joseph Cudjoe, and Eliane Smith as well as my research assistants at FGCU: Monica Piotter and William Mack. I must also thank Caitlin Lavelle, Nicole Grazioso, and Maria Zamora as well as the anonymous referee who reviewed the manuscript. The research for this project was completed with the financial support of the University of Iowa History Department, the Rockefeller Archive Center, and the Fulbright Program administered by the J. William Fulbright Foreign Scholarship Board and the United States Information Agency.

List of Abbreviations

ACD	Uttar Pradesh Agriculture (c) Department
AD	Uttar Pradesh Agriculture Department
BHC	benzene hexachloride (an insecticide)
CD	Uttar Pradesh Colonisation Department
CI	Census of India
CM	Chief Minister
CTO	Central Tractor Organization
DA	Government of India Department of Agriculture (pre-1947)
DDT	dichloro-diphenyl-trichloroethane (an insecticide)
FD	Uttar Pradesh Forest Department
FPB	Uttar Pradesh Food Production (B) Department
FPD	Uttar Pradesh Food Production Department
GOI	Government of India
GOUP	Government of Uttar Pradesh
IUCN	International Union for Conservation of Nature and Natural Resources
MA	Government of India Ministry of Agriculture (post-1947)
MII	Malaria Institute of India
NGO	Non Governmental Organization
PUDR	People's Union for Democratic Rights
PHD	Uttar Pradesh Public Health Department
PWD	Uttar Pradesh Public Works Department
RAD	Uttar Pradesh Revenue (a) Department
RCD	Uttar Pradesh Revenue (c) Department
RD	Uttar Pradesh Revenue Department
RRD	Uttar Pradesh Relief and Rehabilitation Department
TBDC	Tarai and Bhabar Development Committee
TBGE	Tarai and Bhabar Government Estates
TD	Uttar Pradesh Transportation Department
UP	(before 1947) United Provinces (after 1947) Uttar Pradesh
USTCM	United States Technical Cooperation Mission to India
WHO	World Health Organization
WWF	World Wildlife Fund/Worldwide Fund for Nature

• CHAPTER ONE •

Introduction

This is a study of an increasingly important part of the Indian landscape: the tarai region of the states of Uttar Pradesh [UP] and Uttaranchal. It examines the social process of accelerated land use in the tarai as it has been affected by political, economic, and epidemiological factors and it pays particular attention to the shifting representations of the tarai landscape in the districts of Nainital (later Udham Singh Nagar district), Pilibhit, and Kheri since these contiguous districts have undergone different levels of development.

As a contribution to the literature of the environmental history of India, this study examines the questions of agricultural colonization, wildlife conservation and disease control. The tarai is particularly significant in terms of the difficulties experienced in reconciling agricultural development with the preservation of wildlife.

In 1947, at the time of India's independence, the tarai was a distinctive and unusual part of the Indian landscape. It was primarily a land of swamps, forests, and grasslands and was known as the home of tigers and malarious mosquitoes. By 1975, however, it had been successfully transformed through the interaction of many actors into a normal or usual part of the northern Indian landscape. The process of transformation can be seen as one of normalization, that is, the tarai was transformed into a normal or usual part of India. In terms of public health, wildlife conservation, and agriculture, the tarai differed little from other parts of India. The present day tarai can be interpreted as having been integrated, more or less fully, into post-colonial India.

This study will contribute to the growing literature on the environmental history of India. It will mark a departure from most such works, which focus on the role of colonial and postcolonial forestry and forest control, and will examine wider questions of agricultural colonization, wildlife conservation and disease control. The tarai is particularly significant in terms of the difficulties experienced in reconciling agricultural development with the preservation of wildlife. Furthermore, the entire process of land reclamation was made possible only by the success of various malaria control programs.

Finally, this project will suggest some new lines of inquiry in its effort to integrate the analysis of these themes—namely agricultural development, disease control, wildlife conservation, and forest extraction.

By 1975, the tarai had been successfully transformed into a normal or usual part of the Indian landscape. In terms of public health, wildlife conservation, and the green revolution, the tarai differed little from other parts of India. Firstly, after the National Malaria Eradication Programme had run its initial course, the problem of malaria resurgence in the tarai was no different from similar resurgences in many parts of India, such as Maharashtra. Secondly, when the Wildlife (Protection) Act was passed in 1972, wildlife conservation became a concern of the GOI and the issues, such as villager access to forest resources and grabbing of forest land surrounding Corbett National Park and Dudhwa National Park, were no different than the issues surrounding, for example, Ranthambore National Park in Rajasthan. Lastly, the successes, such as increased production, and problems, such as social dislocation and exhaustion of the soil, of the "green revolution" were similar in the tarai to those in other parts of UP, Haryana, and the Punjab. After 1975, the changes in these three areas in the tarai were similar to the changes elsewhere in India so that the tarai ceased to be an unusual or exotic part of India. In other words, the present day tarai can be interpreted as having been integrated, more or less fully, into post-colonial India. This study proposes to reveal how this was done.

Landscape, History, and Culture

In examining the transformation of the tarai, one focus of this study will be on the agricultural colonization (reclamation and settlement of cultivators on the land) of the tarai after 1947. Reclamation and settlement operations are of primary interest but this study will also involve a secondary theme—a study of the shifting and varied representations of the landscape. These various representations are important because they both guided reclamation and in turn fed back upon and informed the normalization of the tarai. The term "landscape" denotes more than the scenery or even ecosystem and refers to wider idea of a *cultural landscape*. In geographical-historical studies there are two types of landscapes: natural and cultural.[1] Ideally, a natural landscape is an arbitrarily-bounded area that is free from human influence. In more functional terms, a natural landscape is a given area that is perceived to be free (or relatively free) from human influence. Of course, a perception that an area has not been influenced by humans can be faulty—the area may have been indirectly influenced by humans or, as was the case with 19th century tarai where agricultural land had reverted to forest, the direct impact of humans may be easily overlooked.² Errors of this sort would enable a viewer to perceive a cultural landscape as a natural one.

A *cultural landscape* is an arbitrarily-bounded area with a recognized human influence. As geographer James Duncan puts it: landscape "is a culturally produced model of how the environment should look."³ This influence, however, is not one-sided, because the cultural landscape is produced by the interaction of a dynamic nature and humans organized in the form of culture groups. It is important to note that the human side of this interaction is never passive or accidental; all human groups consciously change the surrounding environment to some extent, whatever their technological resources.⁴ Even small earthen dams and forest fires can effectively alter the environment, albeit on a limited scale.⁵

The concept of a cultural landscape originated in the work of Carl O. Sauer and Paul Vidal de la Blache and has been adopted by Annalistes like Lucian Febvre and Fernand Braudel and environmental historians like Donald Worster and William Cronon.⁶ Vidal is known as the founder of the school of "la géographie humaine" which includes a criticism and rejection of environmental determinism—that is, the concept that a given terrain, weather, or flora and fauna inflexibly controls or determines human social and economic organization.⁷ He argued that humans and nature were so entangled that it is difficult to distinguish the influence of one on the other.⁸ Vidal, however, did not use the terms "culture" or "nature" but referred to "genres de vie" (modes of life) and "milieu" (setting) respectively.⁹ Vidal described the links between humans and nature as an ongoing dialectic or dialogue between milieu and genres de vie. he argued that while humans responded to changes in the milieu or environment, they also act to alter the physical environment.¹⁰ Geographer William Norton has described the Vidalian approach as seeing land as a frame for human activity.¹¹ Thus the physical environment serves as a context that permits some human activities, but prevents others.

Carl O. Sauer is known as the founder of the "landscape school" which bears similarities to the ideas of Vidal including the rejection of environmental determinism.¹² Sauer argued that "[t]he cultural landscape is fashioned from a natural landscape by a culture group. Culture is the agent, the natural area is the medium, the cultural landscape is the result."¹³ Sauer also emphasized the need to historicize cultural landscapes by looking at successive developments and series of changes that produced them."¹⁴

Environmental historians have adopted a similar view of the interaction of humans and nature. William Cronon argues that it is necessary to assume a dynamic and changing relationship between environment and its human users. He describes this as a dialectical relationship where "[e]nvironment may initially shape the range of choices available to a people at a given moment, but then culture reshapes environment in responding to those choices."[15] Cronon further argues that the categories of "environment" and "culture" are not static, but changing. That it, they change not only because of the nature-culture interaction, but they also change due to internal factors independent of the other.[16]

In discussing the interactions between humans and nature it is easy to reify culture and lose sight of concrete groups of people or even individuals or, in other words, to adopt a perspective of cultural determinism.[17] Most environmental historians recognize the necessity of avoiding cultural as well as environmental determinism. As noted above, Cronon sees culture as dynamic and changing while noting how the ideological commitments of individuals can influence how cultural groups interact with the land.[18] Donald Worster points out that in analyzing cultural landscapes it is possible to overemphasize culture. He argues that

> [n]o landscape is completely cultural; all landscapes are the result of interactions between nature and culture. That last point is the crucial one, making environmental history more than social or cultural history, though I agree with Cronon that we should never assume that either nature or culture is an altogether seamless whole.[19]

While there is a dialogue between environment (nature) and culture (genres de vie) there is also a complex dialogue or discourse *within* culture. This dialogue takes place between different social groups and individuals; in the specific case of the tarai it has included the question of whether and how the region is to become a "normal" part of India. The result is a struggle or competition over how nature in the tarai should be perceived, understood, and acted upon.[20]

For a fuller and more nuanced environmental history, it is also necessary to incorporate a discussion of the issues of perception and representation. David Lowenthal and Hugh C. Price note the role of "landscape tastes" in developing the physical landscape in which "[p]eople in any country see their terrain through preferred and accustomed spectacles, and tend to make it over as they see it."[21] More specifically, these idealized images or *a priori* assumptions are the product of a competition of ideas and images as well as a competition for power and resources among groups and individuals.[22]

The term "representation" is used here to refer to the explicit and implicit texts that people make to articulate their varying concepts of the terrain.[23] As such, these texts encapsulate the "preferred landscape." Furthermore, these texts are generated to explain and justify policy decisions or the resistance to them. These representations, and the perceptions on which they are based, guide people's interactions with the terrain.

A central component of the analysis of a cultural landscape is technology. The level of technology available is crucial in realizing or implementing desired changes in the terrain. According to historian Richard White, "The environmental consequences of a technology do not rest like some ghost within the machine…[t]hey are the result of factors and relationships beyond the machine itself."[24] In other words, according to Cronon, "[t]ools and technology are immensely important in shaping natural environments, but their effects are powerfully mediated by the cultures in which they are embedded."[25] Even though certain types of technology may be available in a given area, certain sorts of environmental change are not inevitable.

In the tarai, newly available kinds of technology, such as DDT, radically affected the ability of the UP Colonization Department in the 1950s to transform the physical landscape. Furthermore, as the newer types of technology became available the planners became more ambitious in their plans. The consequences of the new technologies are described by Richards, Haynes, and Hagen:

> By the beginning of the early twentieth century we can safely argue that almost all land in colonial India were "in use." That is, their current condition reflected human decisions to act or not to act to alter that condition. Only the most remote mountainous districts in the Himalayas or the desert might be considered beyond the reach of human intervention. In the past few decades even these exclusions no longer apply.[26]

Thus in the tarai, the introduction of new technology had the effect of making the physical transformation of the land possible and thereby bringing it into use.

Notes

[1] Sauer 1963:333.
[2] Lawrence 1982:14-15 and Sluyter 2002:24.
[3] Duncan 1989:186.
[4] Cronon 1983:13, Leibhardt 1988:24, and Damodaran 2007:127.
[5] Cronon 1993:8.
[6] Braudel 1961:726, Norton 1989:27-44, Simmons 1993:62-65, Williams 1994:9-11, Crosby 1995:1183, Whitehead 1998:30, and Sluyter 2002:7.
[7] Vidal 1903:222-223, Ribeiro 1968:654-655, and Crosby 1995:1183.
[8] Norton 1989:35.
[9] Vidal 1911:193-195, Claval and Nardy 1968:98, Buttimer 1978:60, and Norton 1989:35.
[10] Vidal 1903:221.
[11] Norton 1989: 35-36, 59.

[12] Sauer 1963:315-350 and Sluyter 2002:6-7.
[13] Sauer 1963:343. See also Chakrabarti 2007:27.
[14] Sauer 1963:344.
[15] Cronon 1983:13.
[16] Cronon 1983:9-13.
[17] Agnew and Duncan 1981:51.
[18] Cronon 1983:6.
[19] Worster 1990b:1144. See also Sluyter 2002:24.
[20] See Worster 1993:156-170.
[21] Lowenthal and Prince 1965:186. See also Savage 1984:14.
[22] Sluyter 2002:10.
[23] According to Cronon 1992:1375, the texts that are created tend to take the form of narratives. Cronon argues that narrative is "among our most powerful ways of encountering the world, judging our actions within it, and learning to care about its many meanings."
[24] White 1992:94.
[25] Cronon 1993:8.
[26] Richards, Haynes, and Hagen 1985:525.

• CHAPTER TWO •

The Ecology of the Tarai

Within four months of Indian independence, on January 4, 1948, the first bulldozer began clearing an area of trees on the Tarai and Bhabar Government Estate in a sparsely populated region of Nainital district in the state of Uttar Pradesh [UP].[1] This bulldozer and dozens like it faced difficulties, such as the many hidden pools or bogs that could quickly immobilize a machine. The bulldozer drivers, employees of the GOI Ministry of Agriculture's Central Tractor Organization, also faced many hazards, the most important of which was the threat of malaria. Even with the malaria control efforts of the state government and the World Health Organization [WHO] in the area, malaria continued to be a real threat to the health and welfare of the drivers. Despite all of the difficulties, some 30,000 acres were cleared for cultivation over the next few years, and this clearance, along with malaria control programs, marked the beginning of the Tarai Colonization Scheme.[2] This centrally planned development project served as the means for bringing the tarai region of UP into the postcolonial history of India.

As a sub-part of north India's geography, the tarai has been a wholly distinct ecological zone, part of which prior to 2000 made up the southern portion of Nainital and the northern portions of Pilibhit and Kheri districts of the Indian state of Uttar Pradesh. In 1995, the tarai portion of Nainital district was made into a new district called Udham Singh Nagar which was then transferred to the new Indian state of Uttaranchal in 2000. Before clearance and development, the tarai was a marshy and swampy tract of forest and long grass extending along the base of the Himalayas from West Bengal in the east to Pakistan in the west. The north to south width of the tarai in UP and Uttaranchal varies from about two to fifteen miles. The terrain of the tarai is relatively flat with a gentle downward slope from north to south, southeast, and southwest. Poor drainage causes rivers descending from the mountains to divide into numerous springs and small streams that criss-cross the area with the result that it is prone to water logging and flooding.[3] The primary difference between the tarai and the plains to the south is water. In the tarai, the soil contains more moisture, including a higher water table, than in the plains. Geographer L.R. Singh explains:

> The geography of the Tarai is more of hydrography than topography. Water is the keynote in its time and space relationships. Surface water in swamps and morasses, lakes and sluggish streams, underground water in ... springs and [a] high water-table give it a veritable character of a "half solid" and a "half fluid" passive surface which raises an obstacle of pure inertia to human movements.[4]

Because of these characteristics, this region had been situated outside the mainstream of South Asian history for some 250 years prior to the appearance of bulldozers in 1948. Drainage and swamp clearance, however, would bring the tarai into the twentieth century.

The forests of the tarai are comprised of a variety of tree species, most notably sal (*Shorea robusta*), haldu (*Adina cordifolia*), silk cotton or semal (*Bombax ceiba*), and babul or kikar (*Acacia nilotica*), and neem (*Azadirachta indica*). The tarai's extensive grasslands ("phantas") contain over thirty-seven species of grasses as well as a large number of bushes, creepers, climbers, and shrubs.[5]

Various afforestation programs after 1947 have included native species as well as a variety of exotic species, but the emphasis has been on commercially valuable trees. The native species have included mango, leechi, khair, sal, haldu, semal, and shisham. The introduced species included eucalyptus (from Australia), poplar (from England and Italy), and kububul or ipil-ipil (*Leucaina leucocephala*) (from Mexico).[6]

The native animals of the tarai, still found in Corbett National Park and Dudhwa National Park, include many predators, ungulates, monkeys, snakes, and birds. The more notable species include elephants, tigers, leopards, wild pigs, antelope, nilgai, deer (including chital or spotted deer, hog deer, sambar, swamp deer, and barking deer), crocodiles (including gharial and mugger), porcupines, wolves, foxes, langurs, rhesus monkeys, myna birds, jungle fowl, peafowl, larks, quail, bulbuls, drongos, robins, egrets, and cormorants.[7]

Before 1947, this submontane tract was thinly populated and was primarily known to outsiders as the home of wild animals and malarious mosquitoes. When the East India Company conquered the tarai portion of what now comprises Udham Singh Nagar district shortly after 1800, it quickly consigned the tarai to the periphery of its growing empire. The early administration of the area was deemed unsuccessful by the government of the North-Western Province, and in 1861 it was designated an "extra regulation tract." It then became a government estate administered by a superintendent and remained so until 1947. In 1816, the adjacent tarai area of present-day Pilibhit district was awarded to the Company as part of the reparations paid by the Raja of Nepal after losing the Anglo-Gurkha war. The British, however, found it so undesirable that they forced it upon the Nawab of Oudh (Avadh) in place of the repayment of a cash loan.[8] In the *East-India Gazetteer* of 1828, Walter Hamilton noted that the human population of this area consisted of "[a] few straggling villages" due to the "extreme unhealthiness of the tract."[9] The tarai area of present-day Kheri and Pilibhit then became part of British India when the East India Company annexed Oudh in 1856. The tarai portions of both districts eventually became productive, and financially profitable, as managed forests for the United Provinces Forest Department.

The tarai has had a special place in the consciousness of Indians and Europeans as the site of both the hunting exploits and conservation activities of the famous wildlife author, Jim Corbett. Readers of Corbett's *Man-Eaters of Kumaon* (1944) and a series of successors, all of which were best-sellers, were introduced to an exotic land with a dangerous climate and man-eating leopards and tigers. Corbett's tarai was also the home of simple tribals, good-hearted bandits, and salt-of-the-earth Englishmen. Part of Corbett's wide appeal was that his books were based, according to the author, on his decades-long experience in the region. Unlike the legions of European shikaris who briefly visited the tarai on hunting expeditions, Corbett regularly lived there during part of the year.

While many descriptions of the tarai, before and after Corbett's, emphasized the exoticism of wild animals and malarial mosquitoes, the area always had a significant, albeit small, human population. This human population included forest-dependent populations, like the Buxas and Tharus; nomadic herders who migrated between the hills and the tarai, including Gaddis, Gujjars, Bhotiyas, and Kumaunis; and immigrant farmers from Rohilkund and Punjab. Also known as adivasis, the Tharu, Buxa and Bhotiyas have were designated as "scheduled tribes" by the government in 1967.

The Tharus and Buxas were cultivators who practiced shifting or "slash and burn" agriculture along with hunting and herding. To clear land for farming they would set fire to a section of forest and then cultivate the garden for two or three years; then, when the soil was depleted, they would renew the cycle elsewhere.[10] They also constructed earthen dams and bunds on various streams in order to provide a constant source of water for irrigation. They kept cattle for milk, hunted wild animals and birds, and collected a variety of forest produce for food, medicinal, and construction uses.[11] Many of the Buxas in Nainital district practiced shifting agriculture until the 1940s when the British government required them to live in fixed settlements.[12]

Seasonal migrants to the tarai like Gujjars, Bhotiyas, and Kumaunis visited the area during the winter months in order to graze their cattle, while the Gaddis lived in the tarai year-round. These pastoralists created a number of permanent and temporary settlements in the tarai called "khattas," and they negotiated concessions from the Forest Department that varied from settlement to settlement.[13] Immigrants from Rohilkund and the Punjab came to the tarai in search of farmland. Most of these immigrants settled in Nainital district and became tenants of the Tarai and Bhabar Government Estates [TBGE]. Despite the constant movement of immigrants to the tarai, the number of TBGE tenants who left the Estates annually was larger than the number of arrivals. As a consequence, the acreage under cultivation in the tarai declined from 1918 to 1947.[14]

After 1947, the normalization—that is, the integration of the tarai into the rest of India—began in earnest. From one perspective, the colonization of the tarai can be seen as an attempt to increase the food supply of India by adding land to agricultural production. It can also be seen as part of the Congress' nation-building project presided over by Jawaharlal Nehru, the first prime minister of India, and Govind Ballabh Pant, the first chief minister of UP. This integration was to be a process by which the tarai was transformed from an exotic, jungly, and hazardous landscape to a normalized part of India. The agents of this transformation were bulldozer drivers, malaria control workers, and cultivating colonists. More specifically, the normalization of the tarai was to realize the social goal of replicating the grain agricultural economy of the caste Hindu and Muslim society of neighboring UP districts. This is seen in the declaration of the UP government in 1953 that "[t]he Tarai area is now a smiling country with flourishing cultivation and a contented people happy to find new homes. Those who have carried the burden of this scheme so far are justly proud of the achievement."[15]

The question, however, of how to determine when the tarai became a "normal" part of India has never been completely answered. Beginning in the 1960s, the influence, even domination, of Congressmen like Nehru or Pant over the nation-building process diminished. Under Prime Minister Indira Gandhi's direction of the Congress party in the 1970s, new definitions of what comprised a "normal" part of India emerged. After 1961, the tarai was subject as well to three new forms of development and integration including the activities of enterprising capitalist farmers (who owned large mechanized farms), the efforts of a new Agricultural University established at Pantnagar, and the conservationist interventions of Arjan "Billy" Singh and others.

This process, however, has been subject to many influences beyond India. Various foreign governments and NGOs have sought to influence government policy through donations of technical and financial assistance. These foreign interest groups have had divergent and sometimes contradictory goals. The result is that the GOI has had to find ways to reconcile the demands of Indian and foreign groups and organizations who had a stake in the development or normalization of the tarai. In general, the GOI has subordinated the influences of foreign aid groups to its own perceptions of the country's needs. The GOI set its own development goals and accepted aid from donors when appropriate. For example, the GOI promoted the development of improved strains of hybrid seeds and asked the Rockefeller Foundation for technical assistance.[16] In addition, foreign organizations have sought to influence GOI policies through persuasion and offers of aid. International groups, like the IUCN and WWF, along with the US Fish and Wildlife Service, promoted wildlife conservation in the tarai and throughout India. They pressured India to ban hunting of animals like the tiger, and to establish wildlife preservation projects. The WWF promoted tiger conservation so vigorously that some members of the WWF have claimed that the WWF itself "began" or "launched" Project Tiger.[17] The Rockefeller Foundation, the Ford Foundation, the University of Illinois, the US Technical Cooperation Mission (predecessor to US Agency for International Development) supported agricultural development and the "green revolution." They not only provided financial and technical assistance, but also sought to influence GOI policies. The US government, in particular, used its foreign aid contributions to pressure the GOI on policy issues.[18]

The original vision of the tarai's future advanced by G.B. Pant and others in 1948 was one of small-scale agriculture, more or less homogenous with the adjacent, long-occupied lowlands on the Indian Gangetic plain. Fifteen years later, however, the tarai had become a socially and institutionally more complex setting, because colonists other than peasant farmers had made the region their home. For example, refugees from western Punjab and eastern Bengal were unfamiliar with the climate and geography of northern UP in general and the tarai in particular and the government of UP attempted to ease the process of integration of these immigrants. There has been tension, however, between the new immigrants and the existing population, especially the Buxas and Tharus, and Gujjars. In 1961, the tarai included many prosperous mechanized farms; it also included Corbett National Park, several small private forests, and the Uttar Pradesh Agricultural University at

Pantnagar, the first land grant university in India. In 1965 the North Kheri Forest Division, a profitable timber forest along the India-Nepal border, was declared a wildlife sanctuary by Prime Minister Indira Gandhi, and, in 1973, Project Tiger, a major conservation program, was initiated at Corbett National Park; eight other parks in other Indian states were quickly included in the project. In 1977, during the Emergency, the North Kheri Forest Division was designated by Indira Gandhi as Dudhwa National Park, but, despite its large tiger population, it was not made part of Project Tiger until 1988.[19]

Tigers, Fevers

The tarai has long been known for its wild life. Prior to 1947, wild animals were seen by various government officials as a major obstacle to colonization, second only to malaria.[20] One of the issues involved with the normalization of the tarai therefore was the question of the place of wild fauna in a landscape increasingly being filled up with peasant farmers. The present extent and condition of wildlife in the tarai is the product of long debate among the planners in New Delhi and Lucknow as well as the activities of colonists, foresters, hunters, tourists, conservationists, poachers, and land grabbers. Today, the native tiger, leopard, deer, antelope, wild pig, and many species of birds of the tarai are mostly found only in the forests and grasslands of Corbett National Park, Dudhwa National Park, and Kishanpur Wildlife Sanctuary. While wild animals have been occasionally perceived to be a threat to agriculture after 1972, they have also become an attraction for tourism and related economic development.

Until all sport hunting in India was abolished as part of the Wildlife (Protection) Act of 1972, the dense tiger population of the tarai was well known to both hunters and photographers around the world. British soldiers, ICS officers, viceroys, and Indian nobility and later Indian and foreign sportsmen flocked to the tarai for a chance to bag a fierce tiger. UP forest officer M.D. Chaturvedi observed that before the timber demands of World War II led to mass cutting, the forests of Pilibhit district were primarily noted for their "wily and sophisticated tigers."[21]

After 1947, there was a rapid decrease in the wild animal population, including tigers and prey species. It is useful to look closely at the tiger population, because it serves as an index or indicator species. The decrease in the tiger population, and wild animals in general, in the tarai was due to a loss of habitat, licensed hunting, and especially unlicensed hunting or "poaching". The loss of habitat was a byproduct of land reclamation and the transformation of grass lands and forests into farm land.

There were two forms of licensed hunting that affected the wild animal population: sport hunting and crop protection activities. Sport hunting, although it was identified with British imperialism, remained legal after 1947 because, in part, it was a source of revenue for the UP state government.[22] Furthermore, the GOI promoted sport hunting in the tarai as a form of recreation and as a way to learn about India. According to one government pamphlet:

> Part of one's education is to observe how the supple Indian shikari or jungle coolie can slip noiselessly as any animal through the bushes and thorns. He is lithe and supple, whereas many Europeans are not able to so easily bend low and twist the body this way and that.[23]

Crop protection activities included the use of firearms to frighten away or kill herbivores who posed a threat to crops. One particular threat was the wild pig because a family of wild pigs could devastate entire fields virtually overnight. Firearms were also used to frighten away or kill tigers and leopards that threatened live stock or human beings.

In the first decade after independence, the amount of poaching increased dramatically. E.P. Gee explains that it was the ordinary people of India who upon "realizing that it was now they who owned the animals, often went out into the wild places and massacred whatever they could find."[24] In addition, due to the food shortage of the 1940s and 1950s, the meat of deer, antelope, and other wild animals fetched a high price in urban markets. Furthermore, the skins of deer, antelope, and tiger were also profitable commodities.

In 1972 the GOI became actively involved in wildlife conservation with the Wildlife Protection Act and the formation of an Indian Board for Wildlife task force which proposed Project Tiger.[25] It was recognized that conserving the tiger population would protect the forests and wildlife in general since an entire ecosystem would have to be preserved for the tiger to prosper. Project Tiger initially entailed the creation of nine tiger preserves throughout India. In the tarai, Corbett National Park became the first Project Tiger preserve in 1973, and Dudhwa National Park was included in 1988. By 1982 the tiger population in the nine preserves had nearly doubled.[26] The GOI has thus declared that Project Tiger was a success, but it is not clear that this is the case. Conservationist Arjan Singh argues that much of the population increase was due to the migration of tigers across the border from Nepal when the government of Nepal initiated a program to reclaim its own tarai region.[27] Furthermore, there has also been a great deal of controversy amongst conservationists regarding the reliability of the survey methods—so-called pugmark mensuses—used by the GOI. Thus, it is not clear that the tiger population really did increase from 1972 to 1982; and even if the population did increase, it is not known by how much. Many scholars and conservationists argue that the technique of counting tigers by observing tiger pug marks in the forest is fundamentally flawed. In short, the tiger census figures are unreliable.[28]

The tarai was also notorious for its virulent malaria until malaria control began in Nainital district in 1947. Its swampy conditions were the ideal breeding grounds for several species of malarious mosquitoes, and malaria had been the chief obstacle to large-scale colonization.[29] Moreover, the severity of the malaria in the tarai has been reinforced by human activities. which created breeding places, such as faulty irrigation works and obstructions to natural drainage, for malaria-carrying anopheline mosquitoes.[30] In the tarai, poorly designed and poorly maintained canal irrigation works cut across the natural drainage and caused swamps to develop.[31] In 1946, A.N. Das, Assistant Director of Public Health, UP, sought to explain to other government officials why malaria was such a serious problem in the tarai. He noted that the waterlogging of the soil was due in part to the region's high water table and heavy precipitation and that these conditions were made worse by defective irrigation, roadside borrowpits, and haphazard excavation. The result, he concluded, was that "[a]ll these factors combine to make the area the unhealthiest in the province."[32]

Anti-malarial efforts in the late 1940s and 1950s focused on improvements in public health measures and mosquito eradication. Chief among public health improvements were the efforts of mobile anti-malaria squads which conducted blood and spleen tests and provided, at first, paludrine and, later, chloroquine prophylaxis. In addition, tube wells were dug to replace shallow pools as sources of drinking water. Mosquito eradication involved an attack on adult mosquitoes, primarily the spraying of Dichloro-Diphenyl-Trichloroethane [DDT] and Benzene Hexachloride [BHC] on the inside of huts, bungalows, cattle sheds, and other buildings while mosquito breeding grounds were controlled by spraying stagnant pools with Paris Green and kerosene. In addition, tubewells and canals were combined to provide drainage and lower the water table. It should be noted that concerns were raised at the time about the problem of mosquitoes gaining immunity to DDT and BHC, but no one questioned the possibility of health hazards accompanying the chronic presence of DDT and BHC in the living spaces of villagers.[33]

India received technical assistance in its malaria-control projects from the United States and the World Health Organization [WHO] beginning in 1949. The WHO sent several Malaria Control Demonstration Teams to India, and one operated in the Nainital tarai from 1949 to 1952. The Demonstration Team began by conducting a malariometric survey, which included palpation for enlarged spleens, in order to determine the extent of malaria morbidity. Its members also conducted entomological surveys to determine which mosquito species were responsible for spreading malaria. The vector species in the area were *A. fluviatilis* and *A. culicifacies* with *A. minimus* disappearing from the tarai by 1952.[34]

On January 5, 1952, the United States and India signed the first Indo-American Cooperation Agreement. One of its provisions was the US Technical Cooperation Program, which provided US aid to India's National Malaria Control Programme from 1953-1958. After 1958, US aid went to the National Malaria Eradication Programme. US aid included DDT powder, jeeps, trucks, anti-malaria tablets, and books and laboratory equipment for the Malaria Institute of India.[35]

Malaria, however, again became a problem in the tarai in the late 1960s. It is not clear what caused this malaria resurgence, but in 1979 a mosquito survey was taken in Nainital district, and malarious mosquitoes were detected in large numbers. They were found to breed in areas associated with irrigation facilities. More importantly, these mosquitoes were found to be resistant to both DDT and BHC, both of which were still in use.[36] In 1982-83, in the area around Gadarpur, district Nainital, malaria was found to be endemic. Furthermore, *P. falciparum*, the malaria parasite most common around Gadarpur, showed a "decreased sensitivity to chloroquine."[37] In both 1970 and 1977 the malaria eradication program was given greater emphasis by the GOI.[38]

Reclamation and Development

After 1947 the normalization of the tarai began when much of the forest and grasslands were reclaimed and underwent a process of rapid transformation into agricultural land, beginning in Nainital, Bijnor, and Rampur districts and ending in Kheri and Pilibhit districts. Land reclamation in the form of drainage of surface water, deforestation, and deep soil plowing was mostly performed by the Central Tractor Organization [CTO], a department of the GOI Ministry of Agriculture, on a no-profit, no-loss basis. The costs of reclamation were paid by those who later settled on the cleared and drained land, except for demobilized soldiers who were subsidized by the national government. In addition, settlers were expected to eventually pay rent for their land.[39]

The idea of transforming the tarai began simply with the desire of G.B. Pant and others to improve and modernize the Tarai and Bhabar Government

Estates in Nainital district.[40] This desire was quickly modified by three factors between 1944 and 1948. Firstly, in 1944 and 1945, the Indian army demobilized thousands of soldiers who had no civilian occupations to return to and the GOI apprehended that this might lead to social instability as India approached independence. The GOI then asked the provinces (after 1947 they became states) if they had wasteland that could be developed into agricultural sites on which the demobilized soldiers could settle.[41] Several states, including UP, offered the GOI thousands of acres of swamps, forests, and other marginal lands.

Secondly, in response to a nation-wide food shortage in 1944 that continued to some extent into the mid-1950s, the GOI created the "Grow More Food" campaign to increase production through better farming practices, increased irrigation, and enlargement of the area devoted to agricultural production. The UP tarai was one of the areas targeted by the Grow More Food campaign.

Thirdly, there was the problem of refugees after the partition of India on August 15, 1947. Suddenly the GOI had millions of displaced Hindus and Sikhs from Pakistan for whom it had to find shelter and means of livelihood. One solution was to divide the wastelands already set aside for demobilized soldiers between the soldiers and some of the refugees. This, of course, included the UP tarai.[42]

The effect of these three events was to speed up the normalization of the tarai through the agency of a government reclamation and settlement operation that became much larger than originally intended. Once the development process was altered to accommodate these events, some of the actors involved began to question the goal of and even the definition of what constituted a "normal" part of India. The contemporary tarai shaped by the green revolution and Project Tiger differs greatly from the tarai of small holders and a small Hailey National Park envisioned in 1947 by G.B. Pant.

Notes

[1] GOUP, Bureau of Agricultural Information 1953:52.
[2] RD file number 352/50.
[3] Osmaston 1927:2 and Prakash 1979:72.
[4] Singh 1961:49.
[5] See Singh, Bhati and Maheshwari 1979.
[6] See Yadav and Prakash 1969 and Mathur 1957.
[7] See Kumar and Lamba 1985 and De and Spillett 1967.
[8] Atkinson 1980,2:679 and Hamilton 1820,1:440-441.
[9] Hamilton 1828,2:303.
[10] Bedi 1984:17.

[11] Singh 1956:162, Hasan, 1979:30-40 and Maheshwari, Singh and Saha 1981:10-35.
[12] Nag and Burman 1974:5-11.
[13] Rawat 1993:42-43, 105-108 and Hasan 1992:52.
[14] RCD file number 41c/1948, p. 63-112 and GOUP, Revenue Department Proceedings, November, 1916, pp. 7, 32-33.
[15] GOUP, Bureau of Agricultural Information 1953:15.
[16] Lele and Goldsmith 1989:310.
[17] Mountfort 1983:32 and Luoma 1987:63.
[18] Goldsmith 1988:160.
[19] Singh 1993b:204, Sankhala 1993:122 and Ward 1993:131.
[20] GOUP, TBDC 1947:96.
[21] Chaturvedi 1942:138.
[22] Singh 1993a:74, Gee 1962:9 and Chaturvedi 1969:28.
[23] Anon. 1948:33.
[24] Gee 1992:159.
[25] Singh 1993a:74.
[26] Panwar 1982:137 and Sankhala 1993:9. This figure is based upon official Project Tiger census statistics.
[27] Singh 1993a:91, 96.
[28] See Karanth 1987, Eisenberg and Seidensticker 1976, Panwar 1987, Saharia 1978, Tak and Lamba 1978 and Thapar 1989 for discussion of wild animal census methodology.
[29] TBDC 1947:9, Srivastava 1950:151, Chakrabarti 1954:170, Issaris, Rastogi and Ramakrishna 1953:313 and Pampana 1963:428.
[30] Henderson 1949:253.
[31] Rahman, Singh and Pakrasi 1956:157 and Singh 1961:51.
[32] RCD file number 8c/1946, volume 1, p. 77.
[33] GOI, Environmental Hygiene Committee 1949:112-113 praised DDT for being "cheap" and "non-toxic to man and domestic animals" and stated that "It is desirable to apply DDT spray to every house periodically like white-washing." However, M.S. Raghavan did warn that the use of DDT should be limited until research was done on the effects the pesticide had on "useful insects" like the honey bee. He suggested that pyrethrum should be used until the effects of DDT on useful insects was known. Raghavan 1950:35.
[34] PDH file no. 14/1952.
[35] GOI, Ministry of Finance 1959:1, 39-40, 148-149.
[36] Malhotra, Shukla and Sharma 1985:57.
[37] Malhotra, Shukla and Sharma 1985:57-58.
[38] GOI, Directorate of the National Malaria Eradication Programme 1986,1:125.
[39] RRD file no. 1068/48.
[40] RD file no. 63/1946.
[41] RD file no. 26/1946, RD file no. 18/1946, RCD file no. 18/1946, ACD file no. 6/1946, RAD file no. 387/1949, and RCD file no. 1946.
[42] RCD file no. 8c/1946 and ACD file no. 6/1946.

• CHAPTER THREE •

The Production of Exoticism

In the best-selling books of the hunter-turned conservationist Jim Corbett, the tarai region of the United Provinces of Agra and Oudh (UP hereafter) was portrayed as an unusual and exotic land. Corbett described a physical landscape that was very different from the nearby plains of northern India. The tarai was the home of forest-dwelling tribals, free-roaming dacoits (gang-robbers), and beleaguered cultivators. Moreover, the human population of the tarai was only a small part of Corbett's portrait. The real story of the tarai, according to Corbett, was not the presence of human order, however limited, but the vast wildness of nature. The chief residents of Corbett's tarai were the large populations of tigers, leopards, deer, birds, antelope, elephants, crocodiles, snakes, and malarious mosquitoes. Furthermore, this unusual collection of human and animal inhabitants existed between two areas with significant human populations: the heavily populated plains of Rohilkund to the south and the hills of Kumaun, including Nainital town, the summer capital of the United Provinces, to the north. For Corbett's many readers, the tarai was markedly different from either the hills or the plains so that it possessed a reputation—among Europeans and Indians alike—for the unusual and exotic.

Corbett's representation of the tarai was based on both his personal experience of living in the region and on his perceptions of the region. In his books, Corbett described more than the physical landscape of the tarai; that is, his portrayal of the area included a distinct interpretation that constituted what might be called a "preferred" or imaginative landscape. He was not the first or the last to do so.

The preferred landscape of the tarai is crucial to an understanding of the process of the transformation of the tarai's physical landscape or terrain. This is because an interpretation of the tarai's physical landscape includes more than a particularized description of the extant terrain. It also includes a conception of the tarai's potential for transformation. For example, one historical actor might view the tarai as a pestilent wasteland that should be avoided, while another might see vast tracts of fertile soil going to waste. Entailed in these imaginings are distinct courses of action. The former, for example, might advocate that the tarai be left to the animals, while the latter might suggest that it be reclaimed and transformed into productive farmland. In other words, the production of the preferred landscape informs the transformation of the physical landscape; at the same time the extant physical landscape informs the construction of the preferred landscape by revealing the limits or boundaries of a given terrain's capacity for transformation.

As noted above, the tarai had a reputation for being exotic or different from neighboring areas. This reputation rested on a combination of distinctive factors: a relatively sparse human population, a significant wildlife population, and the overwhelming hazard of malaria. The tarai was seen by both Europeans and Indians as possessing a climate that was unhealthy at best and downright deadly at worst. This reputation, however, can be historicized. Jim Corbett was not been the only author of the tarai's preferred landscape. The first European contributions came shortly after the British conquest of Kumaun in 1815. When the British sought to administer the tarai, they quickly developed a preferred landscape in which it was seen both as the home of at least a modest complement of productive peasant farmers as well as an abundance of tigers, deer, and mosquitoes.

Administration of the Tarai

Before 1945, the tarai region was not administered as a single political unit, but was divided among Nainital, Kheri, and Pilibhit districts. Taken as a whole, the tarai was sparsely populated, and it was regarded by Europeans resident nearby and as well as by plains cultivators as virtually empty of human activity. It barely entered consciousness and was known, if known at all, as the home of tigers, other wild game, and malarious mosquitoes. Yet in the 1920s and 1930s, conservationists Jim Corbett and F.W. Champion, through their writings and photographs, had spread the exotic reputation of the tarai throughout India and, ultimately, the world.

Both the physical and preferred landscapes of the tarai reflected the low level of technology available to transform the region. That is, the pre-1945 technology in the region was sufficient for sustained forestry operations and some subsistence agriculture, but was not able to reduce significantly the incidence of malaria, which was the chief barrier to large scale development of the region. Up to 1945, the British and Indians in the tarai possessed quinine, plasmoquine, steam engine plows, tubewells, and improved irrigation and drainage facilities, whereas after 1945 they had new technologies like chloroquine, DDT, BHC, and bulldozers. Indeed, all of these, except the bulldozer, were unavailable anywhere in India prior to 1945.

In other words, pre-war technology was inadequate to fulfill the desire of the Tarai and Bhabar Government Estates [TBGE] administrators to develop the Estates and reshape the physical landscape. It was their ambition to introduce revenue-producing peasants among the forests, wild animals, and mosquitoes; peasants would live in stable, fixed settlements and create what could be described as a patchwork landscape. Aware of the limits of their technology, however, the TBGE administrators encouraged only modest settlement and did not seek to transform the entire tarai.

The British perception of the tarai as an empty and exotic landscape is well-represented by W.H. Moreland of the Indian Civil Service. He observed of the Nainital tarai in 1913 that

> [t]he striking feature of this tract is the difficulty of keeping cultivators on the land, due to the unhealthiness and the unpleasant conditions of life. Fever is very prevalent, the drinking water is often bad, wild animals from the forests have to be reckoned with, markets are few, and social attractions scarcely exist. There is thus always land waiting for cultivators, and any suggestion of the ordinary discomforts sends the cultivators elsewhere, and the land vacated goes out of cultivation and soon returns to the jungle.[1]

Not only did cultivators have to contend with the vagaries of the weather and the pressures of taxes, they also had to cope with malarious mosquitoes, cattle-lifting tigers, and crop-eating elephants, pigs, and deer. Furthermore, while drought was rarely a problem, particularly wet weather could pose numerous difficulties. As Moreland argued

> [t]he chief risk to the prosperity of the tarai is a rise in the water-level following on seasons of excessive rainfall. When this occurs the unhealthiness of the tract increases and the productivity of the land declines: holdings are abandoned, and the coarse grass that springs up harbours predatory animals; the process of deterioration once started is thus cumulative.[2]

The tarai areas of Nainital, Kheri, and Pilibhit districts in 1945 were divided into five different parts: tahsil Nighasan in Kheri district, which included cultivated areas, private forests, and the North Kheri Forest Division; the TBGE of Nainital district; tahsil Puranpur in Pilibhit district, which included cultivated areas, the Pilibhit Forest Division and extensive private forests; the northern portions of Pilibhit tahsil in Pilibhit district, which included cultivated areas and a small portion of the Pilibhit Forest Division; and the *pacca ilaqa* or settled tarai of tahsil Kashipur in western Nainital district.[3] The two forest divisions had been reserved and administered after 1880 by the Forest Department of the United Provinces, while the private forests of Pilibhit and Kheri were owned by local landowners or zamindars.[4]

The settled tarai of tahsil Kashipur in Nainital and the non-forested tarai areas of Kheri and Pilibhit districts were administered by regulations in much the same fashion as neighboring UP districts, while the Tarai and Bhabar Government Estates of Nainital were an "extra regulation tract" in which the Superintendent of the Estates was responsible for all aspects of administration, including judicial, revenue, and public health functions. The TBGE was officially part of the Revenue Department, and the Estates Superintendent was ultimately responsible to the UP Board of Revenue. The population of the TBGE varied over the years, but, after the influenza epidemic of 1918, the population of the Estates fell to its low point in the decades prior to independence. The population in 1911 was 117,761, while in 1931 it was only 87,991.[5]

In 1931, some 6,683 Buxas lived in Nainital district, most of whom were to be found in Bazpur tahsil in the TBGE.[6] In 1946, there were some 50,000 tenants (including many non-adivasis), divided among 1,400 villages in the Estates.[7] Many of the tenants were Tharus and Buxas, but there was a substantial number of immigrants from the UP plains, especially from Rohilkund to the south. The Estates depended on plains immigration to maintain a steady population, yet the Estate's population fluctuated because many immigrants either died of malaria or fled the Estates after they had experienced its poor health conditions.[8] For example, in 1880-1881, settlers originally from the plains fled southward due to a combination of an increased incidence of malaria morbidity and poor rice harvests.[9] It should be noted that Victorian British Estates officials found the climate so objectionable that they resided on them only from December to April and stayed in Nainital town in the Himalayan foothills during the remainder of the year.[10]

The history of the TBGE began in the first half of the nineteenth century with the efforts of the East India Company to develop the tarai in what is now Nainital district. The Sudder Board of Revenue (the chief provincial land-tax collecting body) in 1837 felt that "the tract possesses the advantage of a highly fertile soil and great facility of irrigation..." and should be developed agriculturally.[11] The Board believed that the only barriers to the development of the tarai were the "insalubrity and insecurity" of the region. In constructing their preferred landscape, then, they envisioned the creation of a new physical landscape much like the plains to the south, but with some parts of the swamps and forests being replaced with revenue-producing peasant farms. They argued that the Buxas and Tharus were "honest, simple and industrious in their habits, and...require only the assurance of protection to their persons and property and the aid of moderate advances for food and stocks."[12] In other words, the Board believed that the tribals could be transformed into those revenue-producing peasants of their imagined or preferred landscape. This aid included the distribution of quinine as early as 1855. William Jones, an engineer with the Northwest Provinces government, concluded that this distribution of quinine was relatively successful because the cost of supplying the quinine was more than made up by the increased revenue from healthy cultivators.[13] In addition, British officials relocated some Buxas from particularly swampy areas to grasslands in the southern tarai so that these swamps could be drained. Jones argues that in their new colony the Buxas were able to "improve their material condition."[14]

The Indigenes of the Tarai

The Tharus and Buxas were Hinduized semi-nomadic tribes. The Tharus lived throughout the tarai, but in the TBGE they stayed mostly in the eastern parts of the Estates around the towns of Nanakmata and Khatima, while the Buxas lived mostly in Bazpur tahsil. Both tended to migrate within the boundaries of the Estates and engaged in shifting or "slash and burn" agriculture in which they cultivated rice, pulses, and vegetables. They kept some cattle for milk, but neither tribe consumed beef. In addition to fishing, they hunted such wild animals as deer, antelope, wild pig, porcupine, and various birds.[15]

The Tharus and Buxas, however, were not the only nomadic groups in the tarai. Hill people from higher sub-Himalayan districts, including Bhotiyas, migrated to the Estates during the winter months to graze their cattle. These herders created a number of permanent and temporary settlements in the Estates called "khattas," and had negotiated concessions from the Forest Department that varied from settlement to settlement.[16] The Gujjars, an unscheduled Muslim pastoral tribe, also migrated between the hills and the TBGE.[17]

British efforts to transform the Buxas and Tharus into peasants were not immediately successful. As a result, the government created the Estates in 1863 in order to promote the settlement and development of the Nainital tarai. By this time the government had come to consider the Buxas and Tharus "undependable" and "fickle" and concluded that the tarai could "only be reclaimed by immigration of settlers from the south."[18] The superintendent of the Estates was made responsible for providing all the necessary facilities, such as roads, bridges, and irrigation works, that would attract settlers to the tarai. Attracting settlers was in fact a very important aspect of the superintendent's primary duty, which was to further land reclamation through drainage. The government recognized that "[t]he extension of cultivation, the enhancement of the land revenue, the progress of the people towards a state of prosperity, and the improvement of the now deadly climate, all depend upon the proper application of the water to be drained from the swamps of the Turai."[19] To drain the swamps, the mid-nineteenth century administration focused its efforts on repairing and extending existing irrigation works. Many canals built earlier in the century were in poor repair, even being "silted up and full of weeds."[20] In 1861, British officials emphasized that drainage and proper canal maintenance were necessary to open the tarai to development.[21] At this time, however, they did not think of the swamps as the breeding grounds of malarious mosquitoes, but simply as being "pestilential."[22] It was not until the 1890s that the direct link between malaria and anopheline mosquitoes was established by Ronald Ross.[23]

Well before Ross's demonstration, local officials in UP were aware of the connection between swamps and malaria. Though ignorant of the mosquito's role as the disease vector, William Jones, an irrigation engineer, argued that after the Raja of Rampur had built a dam near the town of Rudrapur, district Nainital in 1821, "a large and noisome swamp has been formed" and that this swamp had caused the ruin of the "once very large and populous village." Jones concluded that "Roodurpoor [Rudrapur] is, I beg to observe, no solitary instance of the ruin of a pergunnah being brought about chiefly by the mismanagement of the dams."[24] Seventy-five years later in 1895, Charles McMinn, unaware of Ross's work, argued that in the tarai of Kheri district the drainage of shallow swamps and ponds "doubtless would extinguish the malaria" endemic in the area.[25] By the end of the century canals in poor repair were understood to cause water-logging and thus to increase the size of mosquito breeding grounds.

That specific locales were a source of disease was a medical commonplace throughout British India in the nineteenth century. It was only after 1900, when the connection between anopheline mosquitoes and malaria had been made, however, that the Estates administration began to take steps specifically designed to combat malaria, such as the creation of medical dispensaries, which made quinine and plasmoquine available to tenants, and the reduction of mosquito breeding grounds through drainage.[26] In 1946-47, the Estates spent Rs. 12,047 on quinine and plasmoquine distribution and Rs. 897 on other anti-malarial activities. During this year, some 27,668 patients received quinine and plasmoquine.[27] These efforts, however, were of minimal effectiveness. Neither quinine or plasmoquine is an effective malaria prophylactic, which means that one can take quinine or plasmoquine and still become infected with malaria. In addition, drainage operations did not have much of an impact on the extent of mosquito breeding grounds, and malaria in the tarai continued to be endemic until the malaria control operations of the Tarai Colonization Scheme.[28] As a result, malaria continued to be the leading cause of death in the Estates during the colonial era. The death rate exceeded the birth rate considerably: from 1914 to 1916 the birth rate was 28.80 per 1000, while the death rate was 51.83 per 1000.[29] Percy Wyndham, Commissioner of Kumaun division, blamed the high death rate on malaria: "malaria as usual is responsible for a very high death toll."[30] Furthermore, malaria was directly responsible for one half of the total mortality and in 1946-47, 1,422 people were reported to have died from malaria in the Estates.[31] It should be noted, however, that other diseases were present in the Estates. In the same year of 1946-47, for example, a cholera epidemic killed 171 people.[32]

Just as the efforts to control malaria were not successful, so attempts to develop the Estates into prosperous farmland failed. In 1916, J.S. Meston, the Lt. Governor of UP, wrote that "in the last 10 years considerable improvements have been effected in theses estates..." and he cited improvements in their physical infrastructure, including irrigation works, roads, and bridges.[33] But "progress" in road construction did not translate into increased agricultural prosperity. For example, the Estates administration tried to improve cultivation through experimentation with different cash crops, such as cotton and sugarcane, the planting of mango and leechi orchards, and through technological innovations like the use of the steam plow. The administrators also established a demonstration farm near Khatima. The Khatima farm conducted dairy farming experiments and also sought to introduce Estates tenants, mostly Tharus, to modern methods in dry farming. The farm also served as a seed store. These agricultural experiments did not meet with good results, however. Cotton proved to be unprofitable, and the Tharus refused to cultivate sugarcane.[34] Furthermore, the amount of land under cultivation did not permanently increase, but instead the acreage under cultivation varied substantially from year to year with a long term small net loss in the acres cultivated.[35]

The TBGE administration recruited settlers from the Punjab, who were given land in villages where labor was in short supply. In 1916, however, J.N.G. Johnson, the administrator of the Estates, lacked "any feeling of confidence in the Punjabis as a permanent resident in the Tarai."[36] He did not elaborate as to why he lacked confidence in the Punjabis, but it may have been the case that the Punjabis were no better in combating the adverse conditions of the tarai than were settlers from Rohilkund.

The protected forests within the Estates were not managed by the Estates administration, but by the UP Forest Department.[37] While these forests contained many commercially valuable tree species, such as *sal, semal, khair*, and *babul*, the quality of the trees was deemed poor by the Department, and much of the wood harvest was used for firewood.[38] The Department, however, did sell *sal, khair*, bamboo, and sandalwood for timber, as well as grass for fodder, firewood, charcoal, and assorted minor forest products.[39] Most sales of forest produce were conducted through public auction, although there were some private negotiations between the Forest Department and contractors. In addition, the forests were open, with certain limitations, to grazing. In 1946-47 some 17,997 cows, buffaloes, sheep, goats, and camels were grazed in the forests.[40]

The Estates also had difficulties with wild animals, which, while contributing to the region's exotic reputation, were a day to day hazard to cultivation. Estates tenants frequently complained about tigers that attacked cattle, but the administration blamed the extent of the problem on the habit of some tenants of letting their cattle roam freely at night.[41] In addition, wild elephants were responsible for crop damage throughout the Estates. The preferred method of removing the elephant menace was to trap and deport the offending elephants.[42]

In contrast, the reserved forest divisions of Pilibhit and Kheri districts were profitable and ran a budget surplus.[43] The higher quality forests in Pilibhit and Kheri had been transferred to government control in 1861 and scientific or "systematic management" was instituted in Kheri in 1886.[44] Sal (*Shorea robusta*) was the most common commercial species, and commercial timber extraction was more extensive (and profitable) in both Pilibhit and Kheri than in Nainital. Nonetheless, these forest divisions were also characterized by an "unhealthy" climate, which limited their exploitation. As in Nainital, the chief culprit was endemic malaria, which limited Forest Department operations to the relatively safer winter months.[45] Only in 1940-42 as the wartime demand for timber grew tremendously, were forestry operations conducted throughout the year. M.D. Chaturvedi, an Imperial Forest Service officer stationed in Pilibhit during this period, stated that forest extraction operations continued while "braving the vagaries of the weather and defying malaria for which the tarai is notorious."[46] Chaturvedi, however, did not discuss the human cost of these operations as measured in malaria morbidity.

It should be noted that the forests in the tarai and, in particular, the North Kheri Forest Division, were subjected to exploitation only after other, more accessible forests in UP and northern India generally had been heavily exploited. As Richard Tucker has argued, British foresters first exploited those forests that were relatively easy to access. They then moved on to less inviting areas only for lack of an alternative.[47] It is clear that this had been the case in UP, especially where there were few plains forests. For example, according to H.R. Nevill, "[t]he forests of Kheri are the most important in Oudh; they not only cover a far greater area, amounting to nearly half the forest land in the province, but also contain superior timber to those of Gonda and Bahraich [districts]."[48] Forest Department officials, like Chaturvedi, found service in the tarai distasteful, but necessary.

Adivasi villages were scattered throughout the reserved forests in Pilibhit and Kheri districts. Tharu villages and lands in the North Kheri Forest Division (in tahsil Nighasan) occupied some 11,238.4 acres of the Forest Division's 188,569.6 acres.[49] Outside the reserved forests of Pilibhit and Kheri there was scattered cultivation by immigrants from the plains and by tribals. In Nighasan tahsil in Kheri, the population, including adivasis within the North Kheri Forest District, was 291,264 in 1931. In addition there was a small population of Gaddis who lived around the village of Maraucha, which is just south of Palia town. The Gaddis were cattle herders until the late 1940s when they took to farming.[50] In Puranpur tahsil in Pilibhit the population in 1931 was 83,460, which included immigrants from the plains as well as adivasis living in the Pilibhit Forest Division. The death rate exceeded the birth rate in both tahsils, and there was an average of 20,000 deaths attributed to "fever" annually in Nighasan. In Puranpur, only 10,542 out of 181,882 cultivable acres were cultivated in 1931.[51]

While the Buxas and Tharus of the Nainital tarai became the tenants of the TBGE, and thus subject to the Estates Superintendent, those living in Pilibhit and Kheri lived in the reserved forests of those districts and interacted primarily with the Forest Department. These Tharus lacked formal tenancy rights and, according to Amir Hasan, "they were left at the mercy of forest officials and [timber] contractors."[52] This interaction, however, was limited, since timber extraction was actively undertaken only seasonally and Forest Department officials resided in the area only during the winter months. The tribals were given rights of limited access to the forest and forest products, such as thatching grass, firewood, timber, and grazing for animals and their interaction with the Forest Department was mostly limited to these contractual transactions. The Forest Department was mostly concerned with minimizing the exploitation of forest products by the tribals. They sought to limit tribal use of forest resources in order to maximize the amount of timber that would be available for sale to private contractors. The Department believed that the exploitation of forest resources by private contractors was economically productive since the contractors purchased the resources at full market value. The adivasis, in contrast, had access to their forest produce allotments at a reduced concession, which limited the profitability of the given forest division. Furthermore, the generation of revenue was one of the primary guiding principles of the Forest Department, and the Department grudgingly allowed adivasis access to the resources granted to them in the National Forest Policy of 1894.[53] The UP Forest Department was proud that the North Kheri Forest Department was consistently profitable throughout the first half of the twentieth century despite the pressures of wild animals, poachers, and adivasis.[54]

In contrast, as tenants of the TBGE, the Buxas and Tharus of Nainital figured prominently in the plans of the Estates administrators. Initially, as noted above, they were meant by the British to become productive revenue-producing peasants. When this did not happen, successive Estates administrators adopted a paternalistic attitude and sought to enlighten and educate "the semi-barbarous Tharoos and Baksas."[55] In the late 1800s, they concentrated on building irrigation works and providing "good huts and drinking water."[56] By the early 1900s, they introduced schools, banks, and medical dispensaries. The Estates administrators even experimented with a sugar factory in the town of Khatima. Although initially suspicious, the adivasis participated in these institutions, some of them more fully than others. For example, the Tharus and Buxas were suspicious of European medicine and were reluctant at first to visit Estates medical dispensaries.[57] The Tharu Bank in the town of Khatima, however, proved attractive as an alternative to traditional moneylenders. The bank even attracted Tharu settlers from Nepal.[58] In 1911, the bank had 470 members.[59] Furthermore, the Tharus did not like to grow sugarcane and had to be pressured by the Estates administration into cultivating it.[60]

W.S. Cassels, Deputy Commissioner, Nainital, stated in 1913 that there was general apathy among TBGE tenants towards education, but that the one exception were "the Tharus, amongst whom education has made great progress in recent years."[61] Tharu schools were successful, in part, because they were staffed by Tharu teachers. Cassels also noted that the Tharus were beginning to use quinine, which he considered to "be a great change considering the well known objection of the Tharus to the use of medicine."[62] It should be noted that the Tharus and Buxas only objected to *European* medicine; they had their own system which was based on medicinal plants found in the forest. For example, the Tharus of Kheri district recognize medicinal properties in some 80 plant species.[63]

Tribal "Immunity" and "Fickleness"

One unusual feature of the relationship between the adivasis and British civil officers was the enduring British belief that the Buxas and Tharus were "immune" to malaria, or at least partially immune. This belief arose from the observation that the adivasis had a lower death rate than the immigrants from the plains. The British believed that if the adivasis survived exposure to malaria in childhood then they became immune to it as adults.[64]

W. Jones in 1855 referred to malaria by the term "Terai fever."[65] He noted that the Buxas "appear exempt from the consequences of malaria."[66] Furthermore, he argued that "Terai fever" was the worst form of malaria and that only Buxas could safely sleep in the most malarial parts of the tarai.[67] H.G. Ross, Commissioner of Kumaun Division, in 1885, stated that "the death-rate shows how unhealthy the climate is [in the TBGE]. None but the Tharus can really stand it."[68] W. Crooke in 1896 differentiated between the Tharus and Buxas. He stated that the Buxas were "proof against malaria," while the Tharus were "generally believed" to be "proof against malaria."[69] He also noted that the Tharus were also "unusually short lived" and that they were remarkable for their "indolence, aversion to service, and incapacity for sustained field labour."[70] The belief that the Tharus were lazy may have originated from their unwillingness to be transformed into revenue-producing peasants living in fixed settlements. In 1921, C.A. Mumford, Deputy Commissioner, Nainital, hedging somewhat, stated that "the Tharus are to some extent immune against malaria and flourish in a tract where neither hillmen nor plainsmen can live." He also noted that settlers from the plains lived in the tarai "at the cost of their health and energy."[71] The idea that the tribals were "immune" to malaria never quite vanished. In 1946, W.W. Finlay, Deputy Commissioner in Charge, Kumaun Division, again stated without qualification that the Tharus and Buxas were "immune" to malaria.[72]

While it is true that malaria morbidity rates were lower among adivasi adults than adults from the plains, this may have been due to certain cultural practices rather than any "racial" differences. The Tharus and Buxas built houses that were better suited for the rainy climate, and they also organized their villages in such a way as to avoid creating mosquito breeding grounds within the village. H.J. Boas, Assistant Commissioner in charge of the Tarai, observed in 1896 that the Tharus

> have adapted their customs and mode of life to the Tarai. They are good and careful cultivators....Their houses are models of cleanliness, are well apart from one another, and are built in such a manner as to prevent the damp accumulating in the walls....The Tharu from long experience of these malarious jungle tracts, has found out what manner of habitation is likely to enable him to withstand the climate, and it is to this that he owes his immunity from fever.[73]

J.C. Robertson, an Imperial Medical Service officer, was impressed when he visited Tharu villages in 1908 and 1909: "the whole appearance of the village gives one an idea of cleanness and prosperity which is very unusual in India."[74] Geographer L.R. Singh argues that Tharu villages were designed to "be safe from water-logging."[75]

Nonetheless, the Tharus and Buxas did indeed suffer from malaria as measured by spleen tests. Robertson observed that among adults of twelve years of age and older, 16 percent of the Tharus and 22 percent of the Buxas had enlarged spleens. In comparison, 82 percent of settlers from the plains had enlarged spleens.[76] While it was customary for British officials, such as Crooke, to dismiss various Indian groups as lazy or indolent, weakness of body is in fact one of the symptoms of malarial infection. The short life span of the Tharus could also have been a result of malaria. If, as Mumford observed, malaria caused settlers from the plains to suffer from a lack of energy, it is entirely possible that it affected the Tharus and Buxas in the same way, albeit to a lesser extent. For example, E.E. Grigg, Officiating Commissioner, Kumaun Division, denounced the Tharus in 1895 for being "as indolent and indifferent a tiller of the soil as it is possible to find anywhere."[77] He also reported that the Tharus claimed that they had to hire field laborers because they were "too weak" to harvest their crops themselves. Grigg's hostile attitude depended in part on his belief that the Tharus were immune to malaria and thus had to be dismissed as lazy or indolent instead of being recognized as chronically ill.

In any case, while the reason for the lower incidence of malaria morbidity among the tribals is unclear, their enduring presence in the tarai posed certain political dilemmas for British administrators. Although the Tharus and Buxas were tenants of the Estates and nominally subject to Estates authority, they could leave the Estates whenever they found residence there objectionable. Nepal was a logical destination for migration because the Tharu and Buxa populations resided, then as now, on both sides of the India-Nepal border. Families or groups of families chose and still choose to live in either country, according to their economic interests. Hence, the British always regarded Tharus as "unreliable" and "fickle" not only because they engaged in mobile "slash and burn" agriculture, but also because they would flee into Nepal during periodic famines and epidemics.[78]

With every administrative change in the Estates, the administrators had to consider the effects on the adivasis or find reasons to shunt them aside. It is notable that the administrators regularly developed separate programs for the them that distinguished them from immigrants from the plains. An example is the Tharu Bank of Khatima, which, as its name implies, was targeted at the Tharu population. At the same time the administrators provided separate credit facilities to tenants from the plains or from the Punjab, such as the cooperative bank in Haldwani. It was not until the 1940s that the Buxas, mostly in Bazpur tahsil, were finally forced by the Estates administration to live permanently in fixed settlements.[79] The Estates administrators may have established separate facilities for the adivasis and desi settlers because they felt that the "backward" tribals needed special encouragement to adopt the ways of peasant agriculture. At the same time the administrators needed special programs to lure settlers from the plains to the area, despite the tarai's poor reputation.

Indeed, the TBGE administrators as late as 1946 were eager to attract settlers to the tarai, in part because of a labor shortage. TBGE Superintendent R.S. Bisht wrote in 1946 that

> [i]t is rather sad that the large areas of beautiful cultivable land in the Estates do not attract agriculturalists from adjacent plains and hill districts and the number of outgoing tenants has increased. Adventurists from far off districts and specially from the Punjab applied in large numbers for land in the Estates and in spite of allotments only very few have actually turned up.[80]

The tarai's reputation for unhealthiness may have deterred some of the Punjabis from taking up their allotments.[81]

Shikar and Wild Animals

Although the tarai was already lightly inhabited when first included in British India, it was better known among Europeans as the abode of wild animals. It was this combination of a small human population with extensive forests and many game animals that led to the tarai's exotic reputation. While not as dramatic in its impact on the land as agriculture, shikar was a central part of both the physical and preferred landscapes of the tarai. The British believed that the area provided excellent opportunities for "sport," and this interpretation influenced British developmental activities. For example, roads were built in the tarai to facilitate shikar operations.[82]

By the mid-nineteenth century the tarai had become a popular destination for British sport hunters or shikaris. Right through the end of British rule government officials, aristocrats, including viceroys and soldiers, hunted in its closed forests. Officers who were stationed in the tarai often took advantage of opportunities for shikar, while those stationed elsewhere traveled there to shoot while on leave. European officials throughout India were given hunting and fishing privileges, and in Kumaun certain officials were exempt from paying hunting permit fees.[83]

For example, Percy Wyndham, when Commissioner of Kumaun, took full advantage of his proximity to the tarai. Govind Ram Kala, a UP Revenue Department official, and Jim Corbett both recorded Wyndham's fondness for shikar. Corbett and Wyndham often hunted together, and, according to Kala, Bazpur tahsil in the TBGE "served as a happy hunting ground for Mr. Wyndham and his friends."[84] Furthermore, from the mid-1920s onward, Corbett organized all shooting parties of visiting dignitaries.[85] As late as January, 1947 Corbett arranged the hunting expedition of the Viceroy, Lord Wavell.[86]

Most Indians were not licensed to own guns or to engage in shikar; gun licenses were mainly granted by the government to protect crops from wild animals.[87] The effect was to reserve shikar for Europeans and their friends and allies. In 1919, G.B. Pant charged that villagers in Kumaun were being denied gun licenses so that the number of game animals available for European shikar would be maximized. Pant argued that the villagers needed firearms to protect themselves and their crops from wild animals. Instead, "[t]he children of the soil should feed the wild beasts so that the latter may be fattened to provide game for the strangers, whose title varies directly in accordance with the distance of their homes from the place of shikar."[88] In other words, Pant argued, district officials like Wyndham were manipulating access to the forest in order to maximize shikar opportunities for themselves to the detriment of villagers. Wyndham and other officials interpreted the tarai in such a way as to legitimize European shikar and stigmatize Indian shikar. Thus any Indian who killed a wild animal was guilty of poaching and faced a variety of judicial penalties, which were determined by officials like Wyndham. While there is certainly some truth to Pant's charge, the government in 1919 was probably less concerned with shikar and more concerned with the widespread peasant protests in Kumaun, which involved the burning of reserved and protected forests. After the massacre of a peaceful crowd in Amritsar, the nationalist movement was becoming more active, and officials were reluctant to grant Indians in Kumaun and elsewhere greater access to firearms.[89]

Indeed, Kala observed that after 1919 Wyndham "did not like others shooting in the Tarai [of the TBGE] and the forest officer-in-charge of the Tarai and Bhabar, who was a subordinate of his, rarely issued tiger permits to others."[90] If Kala's observation is accurate, this was Wyndham's personal decision and was not governmental policy as it refers to the denial of tiger permits to other European shikaris. Of course, as a district commissioner, Wyndham himself did not need a permit to hunt.[91]

•The Production of Exoticism• 33

For a century Englishmen wrote countless shikar memoirs that recounted in detail their hunting exploits, and the memoirs of pre-independence shikaris like R.G. Burton and R.D. Mackay about their experiences in the tarai are examples of a widespread kind of writing. Burton emphasized, for example, the need to use elephants while hunting tiger in the dense forests and tall grasses of the tarai; while this was less "sporting" than hunting on foot, it was safer.[92] Mackay also hunted from elephants and traveled with a retinue of servants, including local Indian shikaris, who served as guides, mahouts, cooks, and personal attendants. Mackay noted that in the tarai of Pilibhit "[f]orest officials say that tigers are more at home in these jungles and roar their contentment with life."[93]

Mackay, who hunted in the tarai forests of Pilibhit district and the TBGE, commented on the "unhealthiness" of the region. He wrote that the forests of Bazpur, "until the influence and activities of the World Health Organization spread there, reeked with perniciously outsized anopheles and tiger."[94] Such perceptions legitimized and strengthened the tarai's reputation for the exotic and contributed to the official British preferred landscape of the tarai.

Arjan Singh, one of present-day India's leading tiger conservation experts, has also written about his shikar experiences in the tarai of Kheri district. After Singh began farming there in 1946, he killed many wild animals, especially pigs, in order to protect his crops. He also hunted in the nearby North Kheri Forest Division, where the forest contained several shooting blocks "which could be applied for on a roster basis, and where we were allowed to shoot a specified number of animals and live in one of the forest rest houses."[95]

The New Era of Conservation

The first conservationist measures in India were intended to protect individual species. For example, to protect the tiger, the forest department devised a number rules and regulations that limited the annual "harvest" of animals. These rules included closed seasons, rules against night-time shooting, bag limits, and gun and hunting licensing. This early form of wildlife protection was initiated with the passage of the Wild Birds and Animals Protection Act VIII of 1912 by the GOI. The idea behind the Act was to prevent the killing of specific species of "charismatic" birds and animals under certain conditions. The Act did not provide for the preservation of entire ecosystems, such as a national park, for the preservation of all wildlife.

A second form of conservationism took the approach of creating specific areas, such as game preserves or national parks, where wildlife was to be protected. Hailey National Park (the antecedent to Corbett Park) was created in 1935, but there were several other game sanctuaries located throughout South Asia before 1947.

Recognizing that the enforcement of the Wild Birds and Animals Protection Act of 1912 was inadequate, the GOI, various princely states, several provincial governments, and many environmental NGOs, such as the UP Game Preservation Society and the Delhi Game Preservation Society, held an All-India Conference for the Preservation of Wild Life in January, 1935. Several issues were debated at the Conference and several decisions were made. The Conference identified species such as the Asiatic lion and pangolin that deserved special protection and decided that the Forest Department should be have the primary responsibility for enforcing game laws. The Conference also recommended several revisions or the game laws: prohibition of the sale of the possession, sale or purchase of those birds or animals which are to be protected, outlawing of hunting from motor vehicles, the banning of shooting by artificial light, the banning of hunting near water or salt-licks, restrictions on wildlife photography, and the prohibition of killing fish with poison or explosives. The implementation of these recommendations, however, was delayed by the beginning of World War II.[96]

The tarai serves as a good example of both kinds of conservationism—species-based and area-based—and it provided ample opportunity for wildlife photography as well as shikar. F.W. Champion, an Imperial Forest Service officer, was a pioneer of wildlife photography in India and was active throughout Kumaun. Anticipating by decades the conservationist norms of the present, Champion objected to shikaris who were "destroying the many and beautiful creatures" of the forests.[97] In the 1920s, Champion argued that wildlife photography was possible, despite the unsuccessful attempts of previous photographers. In his book, *With a Camera in Tiger-Land*, first published in 1927, Champion wrote, "I believe this to be the first book ever published which is illustrated throughout with photographs of wild animals, just as they live their every-day lives in the great Indian jungles, away from the ever-destroying hand of man."[98] Jim Corbett, inspired by reading *With a Camera in Tiger-Land*, also took up photography in the tarai in the 1930s. Corbett was especially interested in filming tigers in motion rather than taking still pictures. Like Champion, Corbett argued that "the taking of a good photograph gives far more pleasure to the sportsman than the acquisition of a trophy."[99]

Both Corbett and Champion were shikaris turned conservationists. While neither wrote in detail about his decision to favor photography over hunting, a factor in their decision was their shared perception that the population of game animals in the region was in decline. Champion and Corbett suggested that overhunting by both licensed Europeans and unlicensed Indians was responsible for lowering the game animal populations. For example, Corbett, in a 1932 newspaper article, argued that the balance of nature was being disturbed by the "unrestricted slaughter of game" by shikaris as well as by villagers.[100] Champion wrote in *With a Camera in Tiger-Land* that

> I would close this book with an appeal to others who do not enjoy spilling the blood of beautiful animals, many of which are rapidly being exterminated, to abandon the rifle in favor of the camera, the use of which provides the pleasures and excitements so dear to the heart of the big-game hunter.[101]

Champion argued that there was a 25 per cent decrease in the population of all wild animals in the North Kheri Forest Division due to the increase in hunting in the area and concluded that the Wild Birds and Animals Protection Act of 1912 was a failure primarily because it was unenforceable.[102] Unlike in other parts of India, the tarai after 1918 suffered a decrease in the human population, and the area under cultivation was relatively steady until 1945; neither Corbett nor Champion referred to any loss of wildlife habitat in the tarai, although Corbett was familiar with the loss of habitat in the Kumaun hills, which was the result of forest burning.[103]

In 1936, Corbett along with R.C. Morris and Hasan Abid Jafry founded the magazine *Indian Wild Life*, and in the second issue the magazine published an editorial that argued in favor of wildlife conservation:

> Game Preservation wherever it may be undertaken embodies the same principle—the principle that, in order to afford game animals that peace and protection which will enable them to live and reproduce their kind without damage to man, man should only be allowed to damage them under certain rules and should be restricted from ruthless destruction.[104]

It should be noted that neither Corbett or Champion expressed any personal regret for their own roles in reducing the animal population; having renounced hunting they simply urged others to consider photography and conservationism.

While Corbett, Champion, and other conservationists argued that there was a decline in the game animal population, there is little hard evidence of such a decline. Neither the GOI nor the government of UP had conducted any formal animal censuses in the first half of the century and all animal population estimates were impressionistic. In other words, Corbett argued that the wild animal population *was* in decline because it *seemed* to be so. Not all observers shared the impressions of Corbett and Champion: based upon his experiences in the region, forester H.G. Champion did not notice any scarcity of wild animals. He felt that the forest department's licensing of hunting ensured a stable wildlife population.[105] Arjan Singh, however, has suggested that the information that Corbett described in his books of tigers and leopards killing porcupines, cattle, and even humans was evidence of a decline in the population of prey species, like deer and antelope, during this period.[106]

This phenomenon of declining animal populations, however, was certainly real outside the tarai or even Kumaun. It was a problem throughout India where ever widespread hunting and loss of habitat was common despite gun and hunting licensing.[107] Historian John MacKenzie argues that at this time habitat loss was not the primary cause for a drop in the tiger population but that tigers had become an endangered species by the 1930s through both a high demand for tiger skins and official government policy which labeled tigers as "vermin" and offered a bounty for each dead one.[108]

One result of the perceived decline of wild game populations in UP was the establishment of Hailey National Park in 1935. The park originally comprised 323.75 square kilometers, but was expanded to 523 square kilometers in 1966.[109] Animals resident in the park included crocodiles, karkar, tiger, leopard, elephant, monkey, chital, sambhar, and various species of snakes and birds.[110] While hunting was banned within the park's boundaries, fishing was permitted in specific areas.[111] Moreover, forest operations, including timber extraction, were permitted within the park's boundaries until 1972.[112] Named after the then-Governor of UP, Malcolm Hailey, Hailey National Park was the first national park of India. The park was created after some 20 years of lobbying by conservationists, including Corbett and E.A. Smythies, an Imperial Forest Service officer, who were concerned about what they saw as a decline in game animal populations.[113] There was strong opposition to the creation of the park by hunters, including Wyndham, who objected to the loss of shikar opportunities, but Hailey was the first governor of UP who was receptive to conservationist arguments.[114] Smythies has argued that the location chosen for the park was ideal for conservation, stating that "[h]ere, as nowhere else, nature is unspoilt by contact with man..."[115] Smythies, however, was not entirely correct, because shikaris like Wyndham and Corbett had hunted in the area, and timber contractors had cut timber, on a limited scale. In addition to their lobbying efforts, Corbett and Smythies assisted in the creation of the park by advising the government on the park's boundaries.[116]

The perceived decline of game animal populations represented a major shift in the tarai's preferred landscape, which then influenced a change in the physical landscape. After 1935, with the creation of Hailey National Park, part of the tarai was to be devoted to conservation, a new form of use in which hunting was made illegal. In the 1960s and 1970s even more of the tarai was devoted to conservation uses. More importantly, the park had a profound effect on future efforts to transform the tarai by limiting agriculture and by serving as a reservoir of, and sanctuary for, crop-eating wild animals. Furthermore, Hailey—renamed Corbett National Park in 1957—served as a precedent for the creation of other national parks throughout India and was selected as the first Project Tiger site in 1973.

Corbett was one of the first conservationists in India, but he is most famous as the central figure in his narratives of shikar expeditions and for his wildlife photography. Perhaps unintentionally, Corbett introduced the exotic tarai to a larger audience in India and eventually to the rest of the world through the publication of several works set in the region, beginning with *Man-Eaters of Kumaun*, in 1944. This book was primarily a shikar memoir, but one with a difference: Corbett's main quarry, as the title suggests, was tigers and leopards that attacked and ate humans. That is, the emphasis was on a kind of shikar that benefited terrified villagers whose lives had been disrupted by fear. Most of these adventures occurred in the Kumaun hills, but the tarai figured prominently in *Man-Eaters of Kumaun* and his later books because Corbett not only hunted there, but he "owned" a tarai village there, Choti Haldwani. As the rent-collector of the village, Corbett was essentially a zamindar, although he claimed that he paid the land revenue from his own pocket.

Corbett created for a world-wide readership an image of the tarai as a wild and exotic place on the edge of civilization: it was represented as the home of a few beleaguered villages, wild animals, shikaris, bandits, and endless forests. This image, however, reached the rest of the world just as the process of radical transformation was about to begin. *Man-Eaters of Kumaun* was published in the United States in 1946, only two years before bulldozers appeared in Nainital district to begin land reclamation. After 1945, the UP government, under the leadership of Chief Minister G.B. Pant, imagined a radically different version of the tarai. Pant and others in the UP government saw in the tarai vast tracts of fertile land that were going to waste. They constructed first in imagination a preferred landscape that then drove the massive transformation of the physical landscape that began with the bulldozers in 1948.

Notes

[1] Moreland 1913, Nainital section, p. 2.
[2] Moreland 1913, Nainital section, p. 11.
[3] Pant 1993:224.
[4] These private forests comprised some 64 sq. miles. *Imperial Gazetteer of India*, Provincial Series (1908), volume 1, p. 567. According to Garg 1959:7 the private forests of Pilibhit district were "vested in the State under the UP Zamindari and Land Reform Act on July 1, 1952 as a result of zamindari abolition."
[5] District Gazetteers of the United Provinces of Agra and Oudh, Supplementary Notes and Statistics up to 1931-32 to Volume XXXIV (D) Naini Tal District, p. i; District Gazetteers of the United Provinces of Agra and Oudh, Supplementary Notes and Statistics, Volume XXXIV, Naini Tal District, 1925, p. 1; Supplementary Notes and Statistics to Volume XXXIV of the District Gazetteers of the United Provinces of Agra and Oudh, 1917, p. i.
[6] Nag and Burman 1974:2.
[7] RDC file number 32c/1948, p. 11.
[8] Robertson 1930:112.
[9] North-Western Provinces and Oudh, Proceedings of the Revenue Department, April, 1882, "Report on the Administration of the Tarai District for the Year 1880-1881," p. 2.
[10] Hunter 1885-87,13:207.
[11] Memo by R. Alexander, 1837 in Saksena 1956:230.
[12] Memo by R. Alexander, 1837 in Saksena 1956:230.
[13] Jones 1855:79.
[14] Jones 1855:77.
[15] Singh 1956:162 and Hasan 1979:30-40.
[16] Hasan 1988:9 and Rawat 1993:42-43.
[17] Rawat 1993:105-108 and Hasan 1992:52.
[18] Whalley 1991:158, 271.
[19] Whalley 1991:160.
[20] Robertson 1930:96.
[21] Jones 1855:12-40.
[22] North-Western Provinces, Proceedings of the Revenue Department, March 1861, p. 299.
[23] Harrison 1978:99-100.
[24] Jones 1855:8-9.
[25] McMinn 1895:30.
[26] Chakrabarti 1955:53.
[27] RCD file number 41c/1948, R.S. Bisht, Superintendent TBGE, "Annual Report of the Administration of the Tarai and Bhabar Government Estates, Naini Tal, for the year ending March 31, 1947," p.78.
[28] Pampana 1963:428, Issaris, Rastogi and Ramakrishna 1953:313, and Srivastava 1950:165.
[29] GOUP, Revenue Department Proceedings, November, 1916, memo by P. Wyndham, p. 14.
[30] GOUP, Revenue Department Proceedings, November, 1916, p. 13.
[31] Robertson 1930:112 and RCD file number 41c/1948, p. 6.
[32] RCD file number 41c/1948, p. 74.
[33] GOUP, Revenue Department Proceedings, November, 1916, memo by J.S. Meston, p. 5.

[34] AD file number 283/47, volume 1, p.10 and RCD file number 41c/1948, p. 92.
[35] RCD file number 41c/1948, p. 67.
[36] GOUP, Revenue Department Proceedings, November, 1916, pp. 32-33.
[37] Atkinson 1980:703 and North-Western Provinces and Oudh, Proceedings of the Revenue Department, March, 1901, p. 71.
[38] Osmaston 1927:20.
[39] RD file number 435c, p. 43.
[40] RCD file number 41c/1948, pp. 131-143.
[41] RCD file number 41c/1948, p. 158.
[42] RCD file number 41c/1948, p. 12.
[43] Champion 1933:118.
[44] Sahai 1949:8.
[45] Sahai 1949:3.
[46] Chaturvedi 1942:138.
[47] Tucker 1988:120.
[48] Nevill 1905:9.
[49] FD file number 113AF/1945, volume 2, p. 52.
[50] Singh 1993a:43-44.
[51] District Gazetteers of the United Provinces of Agra and Oudh. Supplementary Notes and Statistics up to 1931-1932 to Volume XLII (D) Kheri District, pp. 15 and i-iv and District Gazetteers of the United Provinces of Agra and Oudh. Supplementary Notes and Statistics up to 1931-32 to Volume XVIII (D) Pilibhit District, pp. i-v.
[52] Hasan 1989:39.
[53] GOI, Ministry of Food and Agriculture 1952, pp. 15, 41-45.
[54] Champion 1933:118 and Sahai 1949:104.
[55] North-Western Provinces, Proceedings of the Revenue Department, memo by H. Ramsay, March, 1861, p. 296.
[56] North-Western Provinces, Proceedings of the Revenue Department, memo by W.H. Lowe, September, 1860, p. 297 and North-Western Provinces and Oudh, Proceedings of the Revenue Department, memo by H. Ramsay, February, 1883, p. 79.
[57] GOUP, Proceedings of the Revenue Department, September, 1911, p. 24.
[58] GOUP, Proceedings of the Revenue Department, September, 1911, p. 20.
[59] Moreland 1913, Nainital section, p. 5.
[60] GOUP, Proceedings of the Revenue Department, March, 1910, p. 152.
[61] GOUP, Proceedings of the Revenue Department, November, 1913, p. 63.
[62] GOUP, Proceedings of the Revenue Department, November, 1913, p. 63.
[63] Maheshwari, Singh and Saha 1981 1981:46-48.
[64] Robertson 1930:112.
[65] Jones 1855:1.
[66] Jones 1855:3.
[67] Jones 1855:15.
[68] North-Western Provinces and Oudh, Proceedings of the Revenue Department, memo by H.G. Ross, October, 1885, p. 33.
[69] Crooke 1896,2:57 and Crooke 1896,4:405.
[70] Crooke 1896,4:381, 405.

[71] Mumford 1921:1.
[72] RCD file number 8c/1946, volume 1, p. 7.
[73] North-Western Provinces and Oudh, Proceedings of the Revenue Department, November, 1896, p. 339.
[74] Robertson 1930:98.
[75] Singh 1956:158.
[76] Robertson 1930:109.
[77] North-Western Provinces and Oudh, Proceedings of the Revenue Department, memo by E.E. Grigg, November, 1895, p. 14.
[78] Singh 1993a:11 suggests another reason why the Tharus were despised by British officials: "They [Tharus] are a simple and honest people, with a delightful sense of fun and a complete disregard for authority."
[79] Nag and Burman 1974:5, 11.
[80] R.S. Bisht, Superintendent of the TBGE, "Annual Report of the Administration of the Tarai and Bhabar Government Estates, Naini Tal for the year ending March 31, 1947" in RDC file number 41c/1948, pp. 63-112.
[81] Both Randhawa 1980-86,4:57 and Singh 1973a:23 discuss the effect of the tarai's reputation on would-be settlers.
[82] Singh 1993a:43.
[83] GOUP, Forest Department Proceedings, July, 1928, p. 21.
[84] Kala 1974:39.
[85] Kala 1974:8.
[86] Wavell 1973:411.
[87] Indian aristocrats were sometimes allowed hunting privileges, usually as part of a pay off made after the loss of their political status and power as rulers of princely states. Other Indian princes had hunting privileges because they owned vast private forests, such as in Pilibhit and Kheri districts. British aristocrats and governmental officials were often invited to these private preserves for hunting expeditions. GOUP, Proceedings of the Forest Department, July, 1928, pp. 19-22 and FD file number 189/1947.
[88] Pant 1993:256.
[89] In 1918, the Arms Act was amended to remove racial discrimination as a consideration for who could possess a firearm. Instead, "arms were readily available to suitable persons; and they were kept from the unsuitable." Robb 1976:46. Pant's objection was that the cultivators who needed firearms to protect their crops were routinely rejected by the government as "unsuitable."
[90] Kala 1974:39.
[91] See UP, Forest Department Proceedings, July, 1928, pp. 19-22.
[92] Burton 1989:178.
[93] Mackay 1968:58-59.
[94] Mackay 1968:69.
[95] Singh 1993a:46.
[96] Chaturvedi 1969:32.
[97] Champion 1990:ix.
[98] Champion 1990:ix.
[99] Corbett 1988:212.
[100] Quoted in Hawkins 1989:216.
[101] Champion 1990:219.

•The Production of Exoticism•

[102] Champion 1934:106, 111.
[103] Booth 1986:181.
[104] Quoted in Booth 1986:182.
[105] Champion 1974:709.
[106] Singh 1993a:83-84.
[107] MacKenzie 1988:182.
[108] MacKenzie 1988:182.
[109] Bedi 1984:15.
[110] Whitaker 1979:38-90 and Smythies 1936:467-471.
[111] Smythies 1936:471.
[112] Gee 1992:92.
[113] Bedi 1984:14 and MacKenzie 1988:289. The Association for the Preservation of Game in the United Provinces [APGUP], which was founded in 1932, was also involved in the creation of Hailey National Park. The APGUP introduced in the UP Legislative Council a bill entitled "UP Wild Life Protection and Preservation Bill." *Indian Forester* 1933: 243.
[114] Bedi 1984:14. Gee 1992:95 quotes from a letter he received from Corbett in 1955, "There was great opposition (from sporting interests) to the formation of the park and as soon as Hailey left, the District Officials combined and reduced the area of the park from 180 to 125 square miles...."
[115] Smythies 1936:469.
[116] See Bedi 1984, Gee 1992, MacKenzie 1988, and Ward 1993.

• CHAPTER FOUR •

G.B. Pant and The U.P. Government

When the government of the United Provinces of Agra and Oudh [UP] began to consider the situation of the tarai in 1946, it did so in the context of proposed improvements to the Tarai and Bhabar Government Estates. At first, the government did not intend to initiate either a massive transformation or normalization of the entire tarai, but only to improve the TBGE's irrigation, transportation, educational, and public health facilities. The government began to consider the status of the entire tarai landscape in light of a combination of events occurring in UP and throughout northern India. By 1947 the government of UP (along with the GOI) was confronted three problems: a long-running food shortage, the rehabilitation of thousands of Punjabi refugees, and the resettlement of demobilized soldiers. The reclamation of the tarai was then seen by the government of UP as one way to address these problems. As the tarai began to draw the attention of government in Lucknow and New Delhi, the area was re-interpreted by officials. No longer seen as useless, they now viewed the tarai as an area with potential. For example, the GOI was interested in finding farmland in the tarai for demobilized soldiers after World War II, while many UP politicians, like G.B. Pant and K.N. Katju, saw the tarai as a way to solve the state's food shortage.

Furthermore, since 1946, new technologies had became available and made large-scale reclamation, even transformation of the land, possible. For example, during World War II DDT had been used by the American and British armies in southeast Asia and had proven to be an effective insecticide; the chemical became available for civilian use at the war's end. In addition, in 1945 and 1946, the American army donated some of its heavy equipment, like bulldozers and tractors, to the Central Tractor Organization [CTO], the land reclamation department of the GOI's Department, later Ministry, of Agriculture.[1] With this equipment, the CTO performed much of the reclamation of land in the tarai in the late 1940s and early 1950s. As these new technologies became available, the desire in government circles for a massive transformation of the tarai grew.

G.B. Pant was the central figure in planning the transformation of the tarai. Pant, the leader of the Congress in UP and the state's chief minister from 1946 to 1954, had a great interest in transforming the tarai and was personally involved in it from 1946 onward. He had long been a critic of the administration of the Estates. As early as 1918, Pant had charged that

> The Tarai and Bhabar Government Estates are under direct and exclusive control of the Government. No zamindars and no tenancy acts intervene between the officers and the tenantry. Yet these estates can beat their rivals, perhaps, only in their insanitary conditions, high death rate, heavy indebtedness, colossal ignorance and extensive illiteracy.[2]

Pant blamed the colonial government for its inept management of the Estates.

In March, 1924, Pant delivered a speech in the UP Legislative Council concerning what he considered to be the gross mismanagement of the Estates by the government. Pant claimed that the tenants of the Estates were in some ways worse off than "the ordinary tenants-at-will in the zamindari estates [in the plains of UP]."[3] TBGE tenants had no security of tenure, and Pant advocated the extension of "full zamindari rights" to the tenants.[4] He argued that the tenants "are most mild and docile" and were "treated in a very autocratic manner. None of them can afford to resist the wishes of even a subordinate local official."[5] Tenants were "required to be abjectly servile and subservient to every person connected with the administration of the estate" and were totally unable to prosper in this chaotic environment.[6] Pant recommended a number of steps to improve the administration of the estates: adoption of modern malariology methods, the creation of demonstration farms to teach local farmers, and the promotion of industry on the Estates.[7] None of these recommendations were adopted at this time.

In March, 1925, Pant gave another speech before the UP Legislative Council in which he charged the government with the continued mismanagement of the TBGE. He argued that the process of revenue collection was faulty in the Estates. He observed that the government collected a net income of Rs. 265,000 from the Estates while at the same time the basic needs of tenants went unmet. He stated that

> I do not at all want that these helpless tenants of the estates should suffer on account of any political considerations. I think that the attention of the Government has not been prominently drawn to the affairs prevailing in the estates and that is mostly the reason that they are labouring under serious disadvantages to this day. The Government cannot possibly cherish the idea that the tenants are in no worse condition than the tenants of other zamindars.[8]

Pant argued that without a regular revenue settlement, the process of revenue collection provided a great deal of hardship to TBGE tenants. Pant, however, failed to persuade the government to alter its methods of collecting revenue in the Estates.[9]

In a second speech in March, 1925 Pant focused on the provision of education and health care in the TBGE. Pant asked the government what its plans were to improve the public health situation in the Estates and was told by the Finance Member S.P. O'Donnell that there was nothing further the government could do to combat malaria in the area.[10]

In a third speech in March, 1925 Pant once again attempted to draw the attention of the government to the nature of the management of the TBGE. Pant stated that "I am told by the tenants of the estates that the present system is harmful to their interests."[11] Pant again suggested a number of reforms to be made to the Estates administration, but none were adopted.[12]

Beginning in 1926, Pant was successful in gaining some reforms in the administration of the TBGE. In July, 1926 Pant made a speech in the UP Legislative Council as part of his effort to extend full tenancy rights to the tenants of the TBGE in which he proposed an amendment that would extend the Agra Tenancy Act to the "villages owned by the Government in the Tarai and Bhabar Government estates" which was passed by the Council.[13]

Also in July, 1926, the UP government passed a special provision of Rs. 23,643 to improve health care in the TBGE. The included the establishment of two traveling dispensaries designed to combat malaria in the Estates. In passing this provision, the government tacitly admitted that it had not paid sufficient attention to the Estates. Pant noted that the government derived a surplus of Rs. 710,000 from the Estates and that the special provision was overdue.[14]

In a speech in August, 1926 Pant proposed an amendment to change the system of revenue settlement in the Estates. Pant stated that "I know that a number of villages have been reduced to a wretched condition, because of the re-settlement being repeated after such short intervals [10 years]." Pant proposed that the settlement be remade every 40 years. The Finance Member O'Donnell argued against the amendment, but the Council passed it.[15]

In March, 1927, in a debate over the TBGE's budget, Pant pressed for a number of reforms in the administration of the Estates, including a reordering of public health programs. Pant argued that

> The net revenue which the Government estates yield is Rs. 225,000, and it is admitted that this is a most unhealthy tract. It is a malarial region subject to attacks of malaria of the most virulent type. It is but fair and proper that the Government should spend money and give some relief to the people from whom taxes are real-

ized So I suggest that a considerable amount should be spent in carrying out measures for the improvement of the drainage, sanitation and for adopting other steps in order to eradicate malaria or at least to minimize the extent or the intensity of its mischief....I also suggest that a much larger amount should be spent over sanitation and drainage."[16]

In reply, O'Donnell stated that the tarai could never be made healthy so the government should concentrate on people in healthy areas. No vote on Pant's motion was taken.[17]

In September, 1928, Pant charged the TBGE administrators with "carelessness and gross negligence."[18] Due to the failure of the monsoon rains in 1928, there was a failure of crops in the tarai with the result of widespread distress among the tenants of the Estates. He argued that

> We have got a very fine system of irrigation there, and the headworks near Kathgodam were put in proper order only a few years ago at considerable cost. Yet all the canals, though they are in perfect order, have not been able to render any service at this important juncture, because silt has been allowed to accumulate there and necessary precautions were not taken in time, from day to day, to keep the system open, continuous, and in such form as to be regularly working.[19]

The tenants petitioned Pant to arrange for relief measures to be taken, but Pant expressed frustration that that he lacked the authority to initiate action. He said that he was reduced to merely reading the petition in the Legislative Assembly. In response, G.B. Lambert, the Finance Member, merely stated that the canals had already been repaired.[20]

By 1946, Pant was in a position to implement major reform in the administration of the Estates.[21] In a June, 1946 letter to the Board of Revenue, B.N. Jha, Secretary to Government, described government policy inspired by Pant:

> I am desired to address you on the subject of preparing schemes for improvements in, and for the welfare of the tenantry of, the Govt. Estates in Kumaun, particularly the T.& B. Estates. The present Govt. is very much interested in improving these conditions. They consider that these Estates should be in a model condition in all matters conducive to material and moral welfare and in particular the matter of health, literacy and agricultural development.

Jha further stated that the government desired to concentrate on improvements in the drinking water supply, anti-malaria efforts, and housing.[22]

Pant had entered public life in the local politics of Kumaun in 1914. By the 1930s, he was the leader of the Congress party in the United Provinces. He served as premier of the United Provinces from 1937 to 1939 and 1946 to

1947 and he served as chief minister of Uttar Pradesh from 1947 to 1954. Pant, however, was more than merely the leader of the Congress party; he dominated politics in UP. He derived his prestige and authority from a combination of his close association with Mohandas K. Gandhi and Jawaharlal Nehru (on more than one occasion Nehru and Pant had adjoining jail cells), his record as a nationalist activist in his own right, and his abilities as an orator and parliamentarian. It is necessary to look at the role of Pant in the government as chief minister and leader of the Congress to understand the state's role in transforming the tarai.

There are at least thirteen biographies of G.B. Pant.[23] All of them share several characteristics. They focus on the public life of Pant as he rose in prominence in the nationalist movement and in provincial and national politics. Before independence Pant served in various local bodies in Kumaun, in the United Provinces legislative council, and as premier of the province. After independence, after serving as chief minister of Uttar Pradesh, Pant served from 1955 to 1962 as Union home minister. All the biographies dwell on the same central themes: Pant's long service in the nationalist movement, his close ties to Nehru and Gandhi, and his skills as an administrator and parliamentarian. Furthermore, several of them include detailed analysis of Pant's major speeches and political decisions. None of them, however, discuss agricultural colonization in UP general or in the tarai in particular. Indeed, agriculture, forestry, and other environmental issues are rarely mentioned at all.

It is important to look at Pant's interest in the TBGE and in agricultural issues to understand the process of land transformation in the tarai. Pant, as noted above, was concerned about the conditions of the TBGE early on in his political career. Pant lived and practiced law for many years in the tarai town of Kashipur, which was only a few miles to the west of the Estates. When the planning process for the improvement of the TBGE began in 1946, Pant was seen in government circles as an expert on the tarai due to his first-hand knowledge and experience of the area.[24] The combination of Pant's perceived expertise and his dominant role in both the UP government and the UP Congress party gave him his enormous influence in transforming the tarai.

Pant first wrote about agricultural issues in 1908 when, as a law school student, he wrote a newspaper article about various ways to improve Indian agriculture. Pant advocated an expansion in canal irrigation, the creation of farmers' cooperatives, reclamation of cultivable wasteland, and the education of farmers in modern agricultural methods. Furthermore, in anticipation of the green revolution, he called for the chemical analysis of farmland soil and the adaptation of new technologies including hybrid seeds and new agricultural machinery. Pant cites the example of "genetic engineering research" in the US that aimed to improve the quality of the wheat crop.[25] He argued that since, according to Plato and Darwin, only the fittest survive, Indians had to adopt modern agricultural methods and technologies or else "before long no trace of us would be left in the world."[26] By combining his interest in modernizing Indian agriculture and in improving the TBGE, Pant laid the foundations for the massive transformation of the tarai.

Other UP government officials shared Pant's desire to develop the tarai. One official argued that if the project were handled in "a really Big Way" the tarai could become an important granary for the state.[27] This official, and others, described the "Big Way" as including large-scale deforestation, extensive fencing (as protection from wild animals), the use of DDT in malaria control, construction of some 40 miles of metalled or paved roads, and mechanized farming.[28] Dr. K.N. Katju, UP Minister of Justice supported the idea of developing the tarai, but warned that reclamation should be done "on a fairly extensive scale. Tinkering with the problem might be merely a waste of money."[29]

There were several tarai improvement plans developed in the 1940s. The plan first plan was written by J.W. Russell, Superintendent of the TBGE, in 1943 as part of the provincial government's effort to plan for post-war reconstruction. Russell concluded that land reclamation was feasible and proposed a plan for mechanized farming on collective farms. He believed that any plan based on attracting new settlers to the tarai in the future was doomed to fail because all past plans that depended on attracting new settlers had failed. He argued that it was necessary to improve the economic condition of the existing tenantry for development of the tarai to succeed.[30] His plan was immediately rejected after Sir William Stampe, Irrigation Adviser to the GOI, condemned the plan as prohibitively expensive.[31]

F.H. Hutchinson, Chief Engineer, UP Public Works Department, Irrigation Branch, in 1946 was asked by G.B. Pant to determine whether the tarai could be developed as part of the Grow More Food campaign.[32] He concluded that the tarai could indeed be agriculturally productive and argued that collective farms were the best way to overcome the difficulties in developing the tarai.[33]

Radha Kant of the Indian Civil Service at the same time as Hutchinson in 1946 investigated whether the tarai could be developed and provide for the settlement of demobilized soldiers and others.[34] He conducted his investigation at the behest of the GOI, which wanted to know "whether by improved drainage a large area in the Tarai can be brought under regular cultivation."[35]

Kant recognized that malaria was the most serious problem in the tarai and argued that malaria control had to be a sustained and on-going effort. Furthermore, he argued that the magnitude of the problems involved in developing the tarai that reclamation "must be tackled in a really BIG way, on a long-term basis in a prosecution of a definite policy of the Government. Half-hearted measures are bound to fail—so the past experience of small and individual efforts tells us."[36]

D.R. Sethi, UP Agricultural Development Commission, developed a "Model Scheme for the Reclamation and Colonization in the Nainital Tarai" in 1946. Sethi and others in the Agricultural Development Commission drew up the plan after reading in the newspaper about the appointment of the Tarai and Bhabar Development Committee [TBDC].[37] Sethi's plan was concerned primarily with the problems of malaria control and land reclamation. In addition, Sethi proposed the creation of settler's cooperative societies. Radha Kant, the member-secretary of the TBDC, believed that this plan was very useful because of its detailed expenditure estimates and made it available to the TBDC members[38].

As both the governmental "expert" on the tarai and the chairman of the TBDC, Pant was very influential even before independence in shaping the UP government's perception of the tarai. Pant believed that the tarai held great promise, but recognized that development of the tract would be difficult. He was, however, enthusiastic about the possibilities of agriculture in the tarai. Pant observed that

> [t]he soil of Tarai is very rich and fertile. Average yield per acre is very high. But there are some hurdles, which have to be got over. It is a malaria-ridden tract and for decades the indigenous population has been dwindling from year to year. But for a more or less ceaseless flow of new settlers from outside the entire region would perhaps have been de-populated and deserted by now....Still the formidable question of drainage will have to be faced. Jungles will have to be cleared and appropriate devices for getting rid of mosquitoes will be called for on a large scale. Model houses of a suitable type will no doubt be designed with due regard to the physical conditions and handicaps. For cultivation greater reliance will have to be placed on mechanical means than on manual labour.[39]

Pant was eager to begin developing the tarai as soon as possible, but he discovered that the process was more complicated than he had initially thought. He stated that

> I was extremely anxious to start agricultural operations on a big scale in order to raise more food and Mr. Hutchinson had been good enough to prepare his scheme at my request...However, I found that it was not possible to bring any land under the plough for the kharif this year [1946].[40]

When Pant realized that immediate action was not possible, he decided that a committee should be created to consider the development of the tarai due to the large cost of the proposed plans devised by Hutchinson and Kant.[41]

Later in 1946, Katju and Pant created the Tarai and Bhabar Development Committee with Pant as the chairman and Katju as the vice-chairman. The committee was asked to consider all aspects of developing the tarai and it also examined the various existing development plans. At the request of Katju, Pant named several people from Kumaun who had personal knowledge and experience of the tarai, but there were no Bhotiya, Gujjar, Tharu or Buxa members named to the committee.[42]

The committee, however, looked beyond the TBGE and examined the possibilities of developing the tarai areas in Kashipur tahsil and Bijnor, Afzalgarh, and Kheri districts as well.[43] The directions or "terms of reference" given to the committee were "to prepare a blueprint of the schemes for the development of [the tarai] for the resettlement of ex-soldiers."[44] This emphasis on the resettlement of ex-soldiers does not mean that Pant's desire to reclaim the tarai to increase the state's agricultural productivity was forgotten. Instead, it represents the GOI's active interest in the tarai. The GOI was very concerned about those soldiers who were displaced from Pakistan while on active duty because many of these soldiers were requesting leave at a time when the government could ill afford it. The government was worried that there would be severe morale problems if the displaced soldiers felt they were being excluded from being able to obtain land in the tarai colonization project.[45] The GOI had offered financial support to any state which offered land to be reclaimed for use by demobilized soldiers.[46] This support took the form of a grant of Rs. 500 for each ex-soldier settled on the reclaimed land, and the government of UP stated that it could not include ex-soldiers in the colonization program without this money.[47] In addition, Katju cited this support as a way to get the reclamation process started. He also observed that "I imagine that along with ex-soldiers we might be able to settle ourselves a lot of non-soldier population."[48] Despite the emphasis on resettling ex-soldiers, then, the reclamation project was seen by the UP government as a provincial development project that included the settlement of ex-soldiers. The government of UP was also very concerned about the shortage of food in the province.[49]

The committee's report was published in 1947 and was accepted by the government in October, 1947 as a basis for land reclamation.[50] In its deliberations, the committee sought to be as comprehensive as possible in considering all the aspects of land reclamation and colonization. The committee began by surveying the tarai landscape of late 1946, including the physical terrain, previous development efforts, and the land tenancy rights of existing residents of the tarai. The committee noted that

> the story of Tarai [sic] is very depressing... [since] life in most of our Tarai villages is one of continual struggle for existence, against the depredations of wild animals, the rank and vigorous Tarai vegetation, the enervating climate, malaria, bad drinking water, high death rate and infant mortality, low birth rate, bad communications and lack of amenities.[51]

Despite this history, however, the committee argued that "the country itself is most attractive, and could be transformed into a beautiful settlement."[52] The committee believed that through cooperative efforts, rather than the individual efforts of the past, the various difficulties, like drainage, could be overcome. Furthermore, the committee stated that reclamation efforts, like the building of roads, houses, and fences as well as anti-malaria operations should begin before colonists arrived in the area with the work being conducted by government labor and at the state government's expense.[53]

The committee drew up plans to eradicate malaria, install tubewells, and build roads. It considered the merits of collective farming, cooperative farming, and a State Farm. The committee also examined the acquisition of land in tarai areas outside of the TBGE as well as the potential property rights of the colonists. The government of UP, through the TBGE, already held the land rights to all the area of the Estates. The occupants of the Estates were tenants who held limited rights to the land, although the Tharus and Buxas had different land rights than those tenants who migrated to the Estates from the UP plains (or the Punjab). Land in the tarai areas outside the Estates, in Kashipur tahsil, North Afzalgarh district, Rampur district, Pilibhit district and Kheri district was held by a variety of people including zamindars, adivasis, the UP Forest Department, and others.

The committee made extensive recommendations on all of these issues. The committee decided that collective farming was not a viable option and recommended that agriculture should be organized along cooperative lines. The committee also made provisions for the creation of one State Farm comprising some 1,000 acres. The committee also recommended the construction of roads and railroads in Kashipur, the TBGE, and Afzalgarh district.

The committee advocated comprehensive anti-malaria efforts. The first and most important action, the committee argued, was to improve drainage in the tarai by closing most of the existing canals, which were seen as a major source of water-logging. It was also necessary to utilize the natural drainage of the region by improving the existing streams and by removing all dams and bunds. Furthermore, the committee argued that while irrigation canals contributed to water-logging, tubewells could serve as a more reliable source of irrigation in addition to being used to lower the water table.[54] Other anti-malaria efforts included removing the heavy over-growth of plants that also served as mosquito breeding grounds. Furthermore, the committee advocated the creation of a permanent anti-malaria organization that would be stationed in the tarai. This organization would focus on control of mosquitoes and protection of humans through the use of larvicides and insecticides in addition to the reduction of breeding grounds. The protection of humans included the distribution of paludrine, mosquito-proof houses, and the proper siting of villages away from mosquito breeding grounds.

The committee also prepared guidelines for determining who would qualify to receive reclaimed land and determined that each colonist would receive 10 acres. The committee also determined the terms on which the land would be given to the settlers, whom it assumed would be male. Firstly, the land would be leased to the settler and the settler would not acquire any proprietary rights. As such the settler could not transfer or divide his holding. Secondly, all settlers were to be required to join a cooperative society and obey the its by-laws. Thirdly, the colonist was to work the land himself and not rent it to a third party.[55] Furthermore, the settler was expected to pay rent as well as repay part of the cost of reclamation and each settler would receive a loan from the government of UP to defray the cost of his house.[56] The committee, however, left open the question of what to do with the existing tenants of the TBGE, which was settled later by G.B. Pant, Radha Kant, and other officials.

The committee also recommended the creation of a separate Board of Development for the Tarai. This board was duly set up in August, 1947 with the Deputy Commissioner of Kumaun as the chairman.[57] Radha Kant, who had been named by G.B. Pant as the Director of the Colonization Department in June, 1947, served as the board's member-secretary.[58]

One other important recommendation made by the committee was to create tree plantations "to provide fuel and small timber for the settlers."[59] Afforestation was necessary because of the extensive deforestation the committee deemed necessary to clear land for cultivation as well as to combat malaria.

The cost of the reclamation of the tarai was enormous. The committee estimated that reclamation and colonization in the area of the TBGE would cost Rs 17,062,900 in its initial stages, while in Kashipur it was estimated to initially cost Rs. 10,192,000.[60] The grand total for the entire project was estimated by the Finance Department to be Rs. 35,982,000. Of this amount, some Rs. 18,000,000 was to be repaid by the colonists who would receive loans from the government of UP.[61]

Despite the support of Pant and others, there was a fair amount of opposition to reclamation of the tarai during and after the deliberations of the TBDC. Committee members Jim Corbett, W.T. Hall, A.P. Watal, and F.H. Hutchinson criticized all or part of the committee's final recommendations. Corbett was concerned about the ecological consequences of the plan's large-scale deforestation. He argued that this deforestation would lead to floods and the reduction of water available in the area's streams.[62] Hall and Watal went further than Corbett and argued that "any removal of forest cover is bound to be attended with grave consequences, increasing the monsoon discharge with consequent loss of soil."[63] Both Hall and Watal felt that any attempt at deforestation would cause large scale sheet erosion which would affect the tarai itself as well as areas downstream. The committee, however, decided that flooding and soil erosion would not be a problem with the use of a system of crop stripping and rotation that would act as "crop cover" in place of "forest cover."[64] Hutchinson was against the planned State Farm and argued that the area of the farm would best be used by allotting it to ex-servicemen, who otherwise would be denied land.[65]

After the report was published, the Finance Department objected to it in May, 1946. The department argued that the cost of the reclamation and colonization project was drastically underestimated by the TBDC. Furthermore, despite the extremely high cost of the project, relatively few people would benefit. Department officials worried that, depending on the level of GOI funding, other programs might have to be cut back in order to avoid "breaking the budget."[66] The department estimated that the plan would have cost the UP government some Rs. 22,000,000 (that would not be repaid by the colonists) for the resettlement of about 7,200 families and argued that the money could be spent more efficiently on other projects.[67] The department concluded that

> these schemes are financially unsound and the happiness and contentment of a few thousand settlers or of the increase in the cultivated area of a few thousand acres is by no means a sufficient compensation for the large expenditure involved. Moreover, the Provincial finances have already started showing signs of strain.[68]

Even though A.N. Jha, Secretary to the Colonization Department, argued that the Finance Department's cost/benefit analysis was faulty, Pant decided that the government could not afford the cost of the entire project. Instead, he decided to begin with the reclamation of the TBGE and to include the tarai areas of Kashipur, Kheri, and Afzalgarh later on. Pant argued in June, 1946 that

> I cannot anticipate the result but the experiment is worth making. We have got thousands and thousands of acres lying waste in the Tarai, while there is a considerable dearth of food and we have to go round the Globe [sic] in search of foodstuffs. If this plan proves successful, we shall be able not only to provide employment for thousands but also to increase the production of vital foodgrains.[69]

It was thus decided that the reclamation project would go forward and the TBDC report was officially approved by the UP government in October, 1947.

The preparations for implementing the plan, however, began before it was formally approved. A.N. Jha began to arrange the logistics of the first stages of the plan in September, 1947 and reported that the Finance Department was very cooperative. Pant was pleased that "[t]hings have begun to move."[70] But as reclamation began in earnest (on January 4, 1948) additional planning had to be done. Firstly, G.B. Pant was persuaded in January, 1948 to include "Harijan" or dalit ex-soldiers and refugees as well as dalit landless laborers from UP in the colonization plan. The government decided that it would be feasible to include dalits because the UP government's Harijan Uplift Board would pay part of the cost. Various departments within the UP government discussed the issue, and by April, 1948 the proposal to include dalits was approved.[71] In addition, in December, 1948, Pant approved a proposal to include graduates of agricultural schools in the colonization program.[72]

Secondly, the government had to decide how deforestation would take place. At issue was the question of the disposal of commercially valuable trees in the TBGE. The Tarai Board of Development and the Forest Department decided that the Forest Department would clear as much of those areas with commercially valuable trees as possible.[73] It was felt that part of the cost of colonization program could be recovered by the sale of the valuable trees. If the Colonization Department and the CTO cleared all the land, the valuable trees would be included with the non-valuable trees and be sold as cheap firewood.

Thirdly, UP's colonization program had to be approved by the GOI before the latter would pay its share of the costs. Generally, the GOI liked the plan, but it thought that the cost of the mosquito-proof houses was too high. The GOI felt that this high cost placed too much of a burden on the colonists who would then be unable to pay back their loans. The government of UP argued that, due to the tarai's special conditions, the cost of houses could not be substantially reduced so it increased the size of the allotment of land from 10 to 15 acres. The government of UP felt that this larger holding would enable colonists to afford their loans.[74] In addition the UP government negotiated a contract with the GOI's Central Tractor Organization to conduct much of the deforestation and heavy plowing. The contract stipulated that the CTO would work on a "no-profit-no-loss" basis. The UP government would pay the CTO's fee, but then recover the money from settlers in the form of long term loans.

By the beginning of 1948, the government of UP, under the influence of G.B. Pant, had come to see the tarai as more than merely a swampy and malarial region with little potential. The government began to view the tarai as a potential granary of UP, if only it could be reclaimed and settled. Furthermore, no one, not even Corbett or Hall, the Chief Conservator of Forests, voiced concern about the consequences of reclamation on wildlife in the tarai. This is due, in part, because neither the TBDC or Pant advocated the reclamation of the land within Hailey National Park. Indeed, during its deliberations, no one on the committee even mentioned the park. In addition, the Forest Department did not object to the concept of reclamation, although it protested vigorously when the Colonization Department accidentally encroached on its land. This is because the land to be reclaimed was, for the most part, not part of the Forest Department's reserved forests. Most of the land of the TBGE was managed by the Revenue Department and the Forest Department's reserved forests in Kheri and Pilibhit were not included in the TBDC report or in subsequent schemes.

It should be noted that as the reclamation of land in the tarai began in January, 1948, the government of UP was already modifying the colonization scheme to allow dalits and agricultural school graduates to participate and receive reclaimed land. The government would further alter the scheme in 1948 and beyond as hundreds of thousands of refugees from Pakistan arrived in UP.[75]

Notes

[1] GOI, Ministry of Information and Broadcasting 1957:8.
[2] Pant 1993:190.
[3] Pant 1994b:17.
[4] Pant 1994b:17. This proposal, however, was never adopted.
[5] Pant 1994b:17.
[6] Pant 1994b:18.
[7] Pant 1994b:20-21.
[8] Pant 1994b:236-237.

[9] Pant 1994b:238.
[10] Pant 1994b:240.
[11] Pant 1994b:241.
[12] Pant 1994b:242.
[13] Pant 1994a:248-250. It is, however, not clear as to how widely this change was implemented in the Estates.
[14] Pant 1995:149.
[15] Pant 1994a:281.
[16] Pant 1995:127.
[17] Pant 1995:128.
[18] Pant 1996:26.
[19] Pant 1996:26.
[20] Pant 1996:27.
[21] It should be noted that Pant's first government appropriated sometime between 1937 and 1939 Rs. 100,000 per year for the improvement and development of the TBGE, but apparently this money was never spent. In 1946, Pant was not certain what happened to it. RCD file number 8c/1946, volume 2, p. 1.
[22] RCD file number 8c/1946 letter from B.N. Jha, Secretary to Government, to K.M. Lal, Secretary, Board of Revenue, June 8, 1946, volume 1, p.1.
[23] There are at least eight biographies in Hindi and five in English including Pandey 1987, Bakshi 1991, Chand 1975, Kashyap and Shah 1988, Neelima 1988, Rastogi 1987, Rau 1981 and Shah 1972.
[24] See RCD file number 26/1946 "Naini Tal Tarai Colonisation Scheme for Ex-Servicemen," pp. 28 and 35.
[25] Pant 1993:109.
[26] Pant 1993:110.
[27] RCD file number 26/1946 "Naini Tal Tarai Colonisation Scheme for Ex-Servicemen," p. 2. Due to the extensive insect and water damage to this document, the name of this official is not known.
[28] RCD file number 26/1946 "Naini Tal Tarai Colonisation Scheme for Ex-Servicemen," pp. 2-13.
[29] RCD file number 26/1946 "Naini Tal Tarai Colonisation Scheme for Ex-Servicemen," p. 28.
[30] TBDC 1947:139 and RCD file number 26/1946 "Nainital Tarai Colonisation Scheme for the Resettlement of Ex-Servicemen," volume 1, pp. 19-30.
[31] TBDC 1947:2.
[32] RCD file number 26/1946 "Nainital Tarai Colonisation Scheme for the Resettlement of Ex-Servicemen," volume 1, pp. 30 and 47 and TBDC 1947:237.
[33] TBDC 1947:237.
[34] RCD file number 26/1946 "Naini Tal Tarai Colonisation Scheme for Ex-Servicemen," p. 47.
[35] TBDC 1947:2.
[36] RCD file number 26/1946 "Naini Tal Tarai Colonisation Scheme for Ex-Servicemen," p. 15.
[37] RD file number 63/1946, p. 3.
[38] RD file number 63/1946, p. 1.
[39] RCD file number 26/1946 "Naini Tal Tarai Colonisation Scheme for Ex-Servicemen," pp. 30-31.
[40] RCD file number 26/1946 "Naini Tal Tarai Colonisation Scheme for Ex-Servicemen," p. 31.
[41] RCD file number 8c/1946, volume 1, p. 39.

⁴² RCD file number 26/1946 "Naini Tal Tarai Colonisation Scheme for Ex-Servicemen," pp. 29-31 and TBDC 1947:3-4.
⁴³ RCD file number 26/1946 "Nainital Tarai Colonisation Scheme for the Resettlement of Ex-Servicemen," volume 1, p. 57 and GOUP, TBDC 1947:2. Pilibhit district was not included.
⁴⁴ GOUP, TBDC 1947:4.
⁴⁵ RAD file number 387/1949, volume 1, p. 1.
⁴⁶ GOI, Department of Agriculture, general section, 1946, file number 32-3/45PL.
⁴⁷ RAD file number 387/1949, volume 1, p. 10.
⁴⁸ RCD file number 26/1946 "Naini Tal Tarai Colonisation Scheme for Ex-Servicemen," p. 28.
⁴⁹ RCD file number 26/1946 "Nainital Tarai Colonisation Scheme for the Resettlement of Ex-Servicemen," p. 154.
⁵⁰ RCD file number 26/1946 "Nainital Tarai Colonisation Scheme for the Resettlement of Ex-Servicemen," p. 153.
⁵¹ TBDC 1947:8.
⁵² GOUP, TBDC 1947:8.
⁵³ GOUP, TBDC 1947:11.
⁵⁴ The committee argued that "tube-wells will not aggravate conditions, as canals do." GOUP, TBDC 1947:17.
⁵⁵ GOUP, TBDC 1947:29.
⁵⁶ GOUP, TBDC 1947:334.
⁵⁷ *The Pioneer*, "Terai & Bhabar Development" August 26, 1947, p. 9.
⁵⁸ The committee also recommended that those governmental officials stationed in the tarai should receive hazardous duty pay consisting of 120 percent of normal pay as well as rent-free housing. GOUP, TBDC 1947:34.
⁵⁹ GOUP, TBDC 1947:33.
⁶⁰ GOUP, TBDC 1947:334, 336.
⁶¹ RCD file number 26/1946 "Naini Tal Colonisation Scheme for Ex-Servicemen," pp. 59 and 72.
⁶² RCD file number 26/1946 "Nainital Tarai Colonisation Scheme for the Resettlement of Ex-Servicemen," volume 1, p. 90.
⁶³ GOUP, TBDC 1947:10.
⁶⁴ GOUP, TBDC 1947:10.
⁶⁵ GOUP, TBDC 1947:47.
⁶⁶ RCD file number 26/1946 "Naini Tal Tarai Colonisation Scheme for Ex-Servicemen," p. 64.
⁶⁷ It may be noticed that the Finance Department's figures do not add up. It estimated that the entire cost of the project would be Rs. 359.82 lakh while Rs. 180 lakh was to be repaid by the colonists and Rs. 220 lakh was not. The remaining Rs. 59.82 lakh are not accounted for in this Finance Department note. However, in a later undated and unsigned note, the Finance Department's estimate was reported as a total of Rs. 440 lakhs. Due to both the rapidly changing situation (regarding refugees from Pakistan) and the haste of the planning and execution of the reclamation and colonization of the tarai, most of the cost estimates were obsolete as soon as they were made. See RCD file number 26/1946 "Naini Tal Tarai Colonisation Scheme for Ex-Servicemen," p. 90.
⁶⁸ RCD file number 26/1946 "Naini Tal Tarai Colonisation Scheme for Ex-Servicemen," p. 73.
⁶⁹ RCD file number 26/1946 "Naini Tal Tarai Colonisation Scheme for Ex-Servicemen," p. 78.
⁷⁰ RCD file number 26/1946 "Naini Tal Tarai Colonisation Scheme for Ex-Servicemen," p. 105.

[71] RCD file number 26/1946 "Naini Tal Tarai Colonisation Scheme for Ex-Servicemen," pp. 107-117.
[72] RCD file number 26/1946 "Naini Tal Tarai Colonisation Scheme for Ex-Servicemen," p. 125.
[73] RCD file number 26/1946 "Naini Tal Tarai Colonisation Scheme for Ex-Servicemen," p. 119. The date of this decision was not included in the file.
[74] RCD file number 26/1946 "Naini Tal Tarai Colonisation Scheme for Ex-Servicemen," pp. 120-125.
[75] See *The Pioneer*, "400,000 Refugees From Pakistan in U.P. So Far" January 6, 1948, p. 1. The figure of 400,000 came from the Chief Parliamentary Secretary, Govind Sahai.

• CHAPTER FIVE •

Realizing the Ideal

The 1946 Report of the Tarai and Bhabar Development Committee was an explication of the UP government's perception and interpretation of the tarai landscape. The government saw the tarai as genuine wasteland, that is land which was not being used productively.[1] G.B. Pant and other officials believed that the tarai could be transformed into thousands of acres of fertile and productive farmland.[2] While this relatively simple and optimistic perception of the tarai was never wholly discarded, the government was forced to continually revise its interpretation of the tarai in the months and years following the report's publication. Not only did unexpected events, such as the arrival in the state of more than 400,000 refugees from Pakistan lead to this re-evaluation, but the government also had to consider the impact of other issues on land reclamation and colonization in the tarai. During the 1950s, for example, the government found it necessary to consider for the first time the intersecting issues of hunting, wildlife conservation, and crop protection. The proximity of Hailey (after 1957, Corbett) National Park to the colonization projects in Nainital district made if difficult for the government beginning to avoid these issues, but the difficulty of reconciling wildlife conservation and maximum resource extraction remained and was not fully resolved at this time.

By the end of the 1950s, the UP government had begun to arrive at a new, more complex perception of the tarai. No longer would there be simply one governmental interpretation of the tarai, because various state departments, like agriculture, public health, and forestry developed different and competing points of view. This divergence of opinion extended beyond the government as different constituencies asserted their interests in the shaping of the tarai. For example, immigrants and settlers began to voice their concerns about the course of development in the tarai. These settlers were unrepresented in the deliberations of the TBDC, and soon after 1950 they began to chafe at the take-it-or-leave-it attitude of the UP Colonization Department. The result of all this contention was that the single, unified vision of Pant, Katju, and the TBDC had faded away by the middle of the 1950s. In the TBDC report, for example, colonists were meant to be smallholders, who worked their own land and were members of cooperative societies. By 1960, however, a group of "enterprising," "progressive," or "capitalist" farmers had emerged who possessed large mechanized farms and were not necessarily members of cooperatives.

Beginning in 1947 the UP government had made a number of logistical arrangements in order for land reclamation operations to begin. Essentially, the government of UP had to create a Colonization Department. Firstly, the government had to fix a headquarters for the tarai colonization scheme officers, which included office space and housing for officials. Secondly, the government hired the Central Tractor Organization to conduct certain deforestation and reclamation operations. The CTO and UP frequently conducted negotiations in order to resolve disputes over CTO costs and fees. Thirdly, the UP Colonization Department had to obtain necessary equipment and supplies. Fourthly, the government had to raise the funds necessary to conduct reclamation efforts. The government of UP eventually received financial and technical assistance from the GOI, the United States, the World Bank, and the World Health Organization [WHO]. Lastly, the government had to make a number practical decisions regarding lines of authority and responsibility within the UP government for the various facets of land reclamation and agricultural colonization. In April, 1948 G.B. Pant decided that it was necessary to clarify the lines of authority in the tarai as several government departments were active in the area. Pant decided that the Director of Colonization, Radha Kant, would supervise and have authority over all government employees active in the tarai, which meant that the Colonization Department gained control over the forests of the tarai.[3]

In creating the bureaucratic and technical infrastructure of the Colonization Department in Nainital district, several tarai colonization scheme officers were stationed in the town of Deoria in Nainital district, including the Deputy Director of Colonization, Harpal Singh Sandhu. The Agricultural Officer, in charge of the Tarai State Farm, Sultan Singh, was stationed in the town of Rudrapur.[4] One problem faced by colonization officers was a shortage of office space in the tarai itself. The shortage was due to an initial lack of buildings available in the area that was to be reclaimed. This shortage, however, was quickly met through construction.[5] In addition, colonization officers and staff were assigned free housing.

Deforestation and Reclamation of Land in the Nainital Tarai

One of the first steps taken by the Colonization Department in the reclamation of the tarai in Nainital was the removal of forest-dependent populations or tribals from the areas to be colonized. The idea was that the reclamation project would be more efficient if the land to be reclaimed was divided into large, compact blocs. The area around Rudrapur was a patchwork of cultivated fields and grasslands occupied primarily by the Buxas which was confiscated by the UP government. The Buxas were given land elsewhere in the tarai, while those Buxas and Tharus who had land in those parts of the tarai that were not included in the colonization area were allowed to keep their land.[6]

The UP government had a paternalistic attitude towards the Buxas and Tharus that reflected somewhat the disdain the former British government held for them. For example, the government, through the UP Bureau of Agricultural Information, echoed the colonial view of the tribals as "immune" to malaria. A Bureau publication stated that "[t]he Tharus and Bhuksas [sic] are the only people who braved the hazards of nature and stuck to the Tarai. Living in neat little hamlets in the midst of the jungle they have acquired almost complete immunity from malaria."[7] Furthermore, this publication discussed the cultural differences between the Tharus and Buxas by focusing on the modes of dress and habits of the women of each tribe. As such, "[t]he Bhuksa women present a contrast to their Tharu sisters in the matters of dress and ornaments."[8] The Bureau observed that Tharu women prefer colorful clothing and ornaments while Buxa women did not. Also, "[t]he Tharu women favour profuse tattooing unlike the Bhuksa damsels."[9] The Bureau, however, did agree with H.J. Boas, J.C. Robertson, and L.R. Singh that adivasi homes and villages were well designed and maintained.[10] Lastly, the Bureau expressed puzzlement that the tribals had achieved some measure of agricultural prosperity. The Bureau noted that the Buxas and Tharus "have, without the benefit of advice from extension workers, evolved the method of eradicating weeds by plowing them under during the monsoon when they are still young before seed formation."[11] Essentially, the UP government's vision for the tarai did not allow for any agency on the part of the Tharus and Buxas.

In April, 1948, the Colonization Department identified roughly 80 square miles (50,000 acres) of land in the TBGE that was to be cleared.[12] About 30,000 acres belonged to the Tarai and Bhabar Forest Division while the remainder was land rented to TBGE tenants.[13] During the deforestation of the area, the Forest Department was given the task of clearing the commercially valuable tree species like semal while the Colonization Department, along with the CTO and the UP State Tractor Organization [STO], deforested those areas covered with non-commercially valuable trees and reclaimed areas with grasslands.[14] They used a variety of techniques to clear the land such as bulldozers to knock over trees and tractors dragging heavy chains to beat down the grass.[15] Then plows would be sent in to turn up the roots of the long grass.[16] The CTO discovered that tree roots posed a particular problem and that the use of root-cutters was ineffective and so relied on the use of heavy disc plows to cut up the roots.[17] Efforts were also made to lower the water table as a way to clear the swamps.[18] In 1948 and 1949, deforestation and reclamation efforts were hindered by a lack of suitable equipment as well as the necessary funds to purchase it. The required equipment was available in the US, but the UP government lacked sufficient hard currency to pay for it.[19] The International Bank for Reconstruction and Development (World Bank) loaned the UP government $1.25 million to solve this problem.[20]

In 1953, the CTO reported that it had felled 31,126 acres of trees and 19,620.5 acres of grassland, and prepared 19,620.5 acres for cultivation by plowing and harrowing.[21] It should be noted that initially the CTO concentrated on plowing up grassland because it lacked the necessary heavy equipment to pursue deforestation which began in earnest in 1949.[22] The Forest Department was given the task of clearing the timber and most of it was used as firewood in the cities of western and central UP.[23] Due to the huge amount of wood to be moved and the lack of interest by private timber contractors, the Forest Department hired laborers from the Gorakhpur Labour Organization to clear the trees.[24]

Anti-malaria Operations in the Nainital Tarai

Anti-malaria operations were an integral part of the reclamation of land in the tarai. Indeed, the anti-malaria part of the reclamation of the tarai began in 1946 before the deforestation part. The first step in anti-malaria operation consisted of the field testing of two new anti-malaria drugs: mepacrine and paludrine in the Nainital tarai. This testing was sponsored by the Malaria Institute of India but it did not result in a clear conclusion. R.S. Srivastava, then Assistant Director of Public Health (Malariology), UP, argued that while paludrine reduced fever in malarial patients more quickly than mepacrine, the effectiveness of paludrine as a malaria treatment was uncertain. He noted that "[i]n an hyperendemic area like that of the Naini Tal Tarai, it is very difficult to evaluate the relapse-preventing properties of paludrine, because of the difficulty to ascertain whether a person had actually relapsed or had been reinfected."[25] Paludrine was eventually replaced by chloroquine as a treatment for malaria.

Intensive anti-malaria or "malaria control" operations by the UP government in the Nainital tarai began in September, 1947. The government set up an "antimalaria unit" in the town of Kichha which included 30 field workers, two insect collectors, and office support staff.[26] This effort included the towns of Kichha, Lalkua, and Rudrapur and covered some 30,000 acres.[27] The area had a population of 2,556 in 1947.[28] The first step of the operation was a survey of malaria morbidity in the area. Malariologists observed that the spleen rate in both children and adults varied generally between 50 and 100 percent.[29] The child spleen index in Kichha was 70 percent, while it was 100 percent in Rudrapur.[30] The adult spleen rate, however, was 40 percent in Kichha and 50 percent in Rudrapur.[31] The next step was to conduct an entomological survey. The most prominent vector was *A. culicifacies*, which constituted 56.2 percent of the anopheline mosquito population in the area.[32] This was a significant finding, because prior to this survey, malariologists had considered *A. minimus* to be the chief malaria vector in the tarai. Malaria control measures included the DDT residual spraying of the interiors of all buildings and paludrine prophylaxis and treatment.[33] These measures had an immediate effect on the extent of malaria morbidity in the area. R.S. Srivastava, now Assistant Director, Malariology, UP, argued that "there has been a noticeable decline in the percentage of malaria cases to total cases in the year 1948."[34]

In 1949 and 1950 the DDT residual spraying of all buildings and paludrine prophylaxis continued. By 1951, however, paludrine prophylaxis was discontinued in Nainital district except on the UP State Farm and in government institutions like the police and schools. Srivastava observed that "the number of malaria cases receiving curative treatment as a routine in 1951 was negligible."[35] The incidence of malaria morbidity in the area continued to decline after 1948 and in 1951, the spleen rate in the area varied between 8.4 and 10.5 percent.[36] *A. minimus* ceased to be a malaria vector while *A. fluviatilis* replaced *A. culicifacies* as the chief vector.[37] By the mid 1950s the UP government began to reinterpret all of the tarai as no longer being a malarial wasteland.

Other anti-malaria measures were conducted by the Colonization Department and the CTO as part of deforestation and land reclamation efforts. Removal of mosquito breeding grounds like surface pools of water (through drainage and the substitution of canal irrigation with tubewell irrigation) and heavy undergrowth reduced the number of malaria carriers.[38] Those surface pools that were not drained were sprayed with Paris green and other larvicides.[39]

A World Health Organization [WHO] Malaria Control Demonstration Team led by P.C. Issaris operated in the Nainital tarai between 1949 and 1952 as compliment to UP's antimalarial units.[40] The team was based in Haldwani and operated in the area between Kashipur and Kichha around the town of Bazpur. There were 14,631 people living in this area in 1949.[41] As with the UP malaria unit, the WHO team began its work by conducting surveys of malaria morbidity and malaria vectors. They observed that the spleen rate in children in the area around Bazpur in 1949 was 94 percent while the chief malaria vectors were *A. fluviatilis* and *A. culicifacies*.⁴² The WHO team also conducted the residual spraying of DDT on the interiors of all buildings in the area. By 1951 there were only 51 reported cases of malaria in the Bazpur area.⁴³ In 1951 the WHO team extended its efforts to the area around the town of Bilaspur.⁴⁴

The UP government's malaria control operations around Kashipur began in 1949 and ended in 1954 and covered about 60 square miles and had a population of 3,808 in 1949.⁴⁵ Here as elsewhere in the tarai, the malaria control techniques used were paludrine prophylaxis and the spraying of DDT on the walls inside all buildings in the area.⁴⁶ As elsewhere in the tarai, the main vectors were *A. culicifacis* and *A. fluviatilis*.⁴⁷ The malaria morbidity rate dropped from 19.4 percent in 1949 to 13.9 percent in 1954.⁴⁸

By 1954, malaria ceased to be endemic in most of the Nainital tarai as the level of malaria morbidity dropped from its original level of 50 to 100 percent to about five percent in Kichha and 14 percent in Kashipur.⁴⁹ In 1955, A.K. Chakrabarti, Assistant Malaria Officer, UP Colonization Department, declared that the tarai colonization scheme "has thus been a success and the heart of a pioneer throbs in sheer delight whenever he pauses to look back upon the stupendous development that was possible chiefly by malaria control in the so-called invincible Terai [sic] of the past. Human effort had once again scored a victory over nature."⁵⁰ Other observers were even more optimistic in their assessment of the anti-malaria operations. The tarai had began to lose its reputation for being an unhealthy, malarial terrain and the UP Agriculture Department in 1950 observed that after three years "malaria had been conquered" in the area around Kichha.⁵¹ In describing the same area the GOI Ministry of Information and Broadcasting declared that "[t]he scourge of malaria has been completely eliminated.⁵² In addition the Reserve Bank of India as early as 1949 stated that malaria in the Nainital tarai had been "eliminated."⁵³ These assessments, of course, were inaccurate. The UP and WHO teams reported that after their malaria control operations the level of malaria morbidity dropped significantly, but they did not claim to have eradicated malaria in the tarai.⁵⁴ Malariologist Diwan Chand offered a more sober assessment of the malaria control efforts in the tarai: "[t]he hitherto notoriously unhealthy tract of [the tarai] have [sic] been converted into a healthier and prosperous agricultural area."⁵⁵

Beginning in 1953-1954, malaria control programs in UP were combined under the Public Health Department and the Nainital tarai was consolidated into two zones of operation.[56] It is important to note that while malaria control efforts were initially successful, complications arose as early as the late 1950s. One such example, was the occurrence of resistance to DDT in mosquitoes so that DDT became increasingly ineffective in killing mosquitoes. In March and April of 1957, the Malaria Institute of India tested the comparative efficacy of diazinon, malathion, and DDT in the Nainital tarai. The tests indicated that while DDT was no longer effective, diazinon and malathion were able to effectively control the mosquito population.[57]

Development of the Nainital Tarai

After the transformation of forest and grass land into arable land, the Colonization Department had to begin the task of building a suitable infrastructure of houses, roads, irrigation works, and drinking water works to attract settlers to the area. As land became available for colonization, the state had thousands of refugees from which to select the settlers who would get land. The Colonization Department, however, initially had great difficulty persuading would-be colonists to take up their land allotments in the tarai. When settlers learned of the tarai's deadly reputation, they would refuse to go there until receiving assurances that the land was safe.

To be eligible to receive reclaimed land in the tarai, refugees had to register with the state. Unofficial refugees were normally denied any chance to receive land, although some army officers who were on active duty and therefore could not register were given special land allotments.[58] Eventually, colonists were drawn from several groups: refugees from Bengal and Punjab, ex-servicemen, landless laborers (including dalits), political sufferers, and graduates of agricultural schools. The selection of settlers and granting of land allotments began shortly after land reclamation operations began and was in progress as early as June, 1948.[59] Most settlers received a plot of 15 acres, although some agricultural school graduates and army officers received 30 or 50-acre plots.[60] To be eligible for land, settlers had to agree to work it themselves and not rent it to a third party.[61] Furthermore, all settlers were required to join cooperative societies as well as contribute Rs. 500 as a deposit for their land allotment.[62] While the GOI paid the deposit on behalf of ex-servicemen, other colonists had to either take out loans or sell their personal effects to raise the money. In the Nainital tarai, settlers were loaned Rs. 3000, which was to be repaid in biannual installments over a 25 year period. Settlers were also expected to pay rent for their land which began in 1949 at Rs.6/8/- per acre annually.[63]. In a pamphlet, the Tarai Farmers' Federation, an NGO based in Nainital district, described the experiences of refugee settlers in Nainital: "Herculean efforts were made by these persons, who invested their life's saving including the ornaments of their women folk and even ran into heavy debt" to purchase farm equipment and pay the deposit.[64] However, as far as the UP government was concerned, setters were defined as male so that land ownership rights did not accrue to any of the women who sacrificed their jewelry.[65]

By July, 1949 the Colonization Department had deforested and plowed some 14,000 acres, of which ex-servicemen had received 500 acres, displaced persons had received 7,500 acres, and political sufferers, diplomats, harijans, local residents and graduates of agricultural colleges had received 3,000 acres.[66] As for the selection of colonists, UP government's Selection Board chose those individuals from the list of displaced people whom it thought would make good agricultural pioneers.[67] The Director of Colonization in 1948 had imposed the requirement that all displaced persons should work on the Tarai State Farm before they received their land grant. The Selection Board then ranked the displaced persons for their ability to perform agricultural work. Settlers who were to receive land were chosen from this list. At first, however, those settlers chosen were not eager to accept their allotments. In 1948, Harpal Singh Sandhu, Deputy Director of Colonization, had difficulty in persuading refugees to take up land in the tarai once they had heard about its deadly reputation. Sandhu noted that the refugees referred to the tarai as "that horrible place" and needed to be reassured that the tarai was safe.[68] Sandhu's method was to begin rice cultivation as a government project as a way to show that agriculture in the tarai was not only possible, but safe. Once the efforts at cultivation by the government and early settlers proved successful, many settlers were attracted to the tarai.[69] Another method the government used to instill confidence in settlers to the tarai was to allot land to eminent persons in selected colonization sites. One example was the allotment of 50 acres to one B.K. Mukerjee, M.L.A. in late 1951. Radha Kant noted that a colony of 500 Bengali families were not coping well with life in the tarai. He argued that

> there is need for the presence of a respectable person of Bengal nationality in the neighborhood to infuse confidence and enthusiasm amongst the displaced persons from East Bengal who are restive in foreign surroundings. A number of families have already left as they are very home sick and the presence of Sri B.K. Mukerjee, who is also interested in jute cultivation, will be of great help both in infusing confidence amongst the settlers from East Bengal and also in the local authorities making use of his services.[70]

In Bengal, Mukerjee had been a zamindar and had lost all of his land during partition, and the UP government felt that since he was both a "respectable" man and a long-time Congress worker he was eligible for a special grant of

50 acres. The government approved his allotment in December, 1951.[71] By 1954 some 1,000 "jute-growing families" from Bengal were settled in the tarai. There was a shortage of jute in India at this time and the UP government hoped that Bengalis, who were accustomed to growing it in Bengal, would take up jute cultivation in the tarai. [72]

By May, 1950, however, there was not enough reclaimed land to give to all those who wanted it, as by this time a very large number of settlers had been persuaded that the tarai was no longer deadly. The UP government then decided to reduce the average grant of land from 15 acres to ten.[73] The UP government also decided in July, 1950 to concentrate on finding land for the UP landless agricultural laborer population as well as UP political sufferers.[74] In supporting this policy, Radha Kant argued that there was no shortage of political suffers, agricultural graduates, and landless laborers from UP who were in need of land.[75] Due to an acute shortage of land in Punjab state, the government of Punjab had contacted the UP government about finding land for political sufferers from their state. The UP government replied that it could not comply for two reasons: there were already thousands of unregistered displaced persons in the tarai for whom land was not available, and the UP government could not afford to financially support the Punjabi political sufferers. Furthermore, if the Punjabi political sufferers were given land, the UP government feared that this would attract an almost endless number of political sufferers and displaced persons from throughout all of India.[76]

After deforestation, reclamation and distribution of land, the Colonization Department embarked on a project of building roads, houses, tubewells, irrigation works, schools and other facilities. By December, 1952, the government had built 38 miles of paved (metalled) roads and 43 miles of unpaved (unmetalled) roads as well as 43 tubewells, which were used to supply water for drinking and irrigation. The town of Rudrapur was selected as an administrative center and there the government built a full service hospital and a high school.[77] Also in Rudrapur, the government built an electricity generating station and set up a series of power lines to electrify the entire colonization area.[78] The government also built some twenty miles of barbed wire fencing to keep out wild animals.[79] Bus service between the towns of Kichha, Rudrapur, and Kashipur was established by the UP Transportation Department in June, 1953.[80]

By December, 1952, 1,289 displaced families from Punjab, 642 families of political sufferers, 173 families of ex-servicemen, 500 displaced families from Bengal, and nine families of agriculture graduates received land in the Nainital tarai.[81] The amount of land reclaimed by the CTO, the STO, and private colonists in the Nainital tarai by June, 1952 was 45,000 acres.[82]

The Tarai State Farm and Cooperative Farming

The Tarai State Farm was established in 1948 by the Colonization Department near the town of Rudrapur in Nainital district. The 16,000 acre farm first served as a demonstration site to show prospective settlers that agriculture in the tarai was both safe and productive.[83] The State Farm also served as a site for agricultural experimentation and as a production site for hybrid seed and purebred poultry, cattle and buffaloes.[84] The Colonization Department also included in the farm a dairy facility and a 1000-acre orchard, planted with mango, guava, papaya, citrus, and banana trees.[85] The fruit surplus from the State Farm (and the tarai generally) was processed at a fruit processing plant built in the tarai town of Phoolbagh.[86] During the year 1951-1952, the farm produced crops of rice, milo, bajra, wheat, barley, oats, and gram.[87] During the year 1955-1956, the farm produced roughly 650,000 pounds of milk, 50,000 maunds of grain, 90,000 maunds of rice seed, and 86,000 maunds of sugarcane seed.[88]

As noted above, all colonists in the Nainital tarai were required to join a cooperative society. After independence, the GOI as well as the government of UP promoted the establishment of farming cooperatives as a way of compensating for the large number of small holdings. Small holdings were held to be uneconomic and the remedy was agricultural cooperatives where farmers could pool their resources to improve the productivity of their small farms.[89] In the Nainital tarai, the UP government hoped that the cooperatives would help in blending the settlers into a new society. The UP government required settlers to join cooperatives, or what it called "land settlement societies" and advanced loans to the cooperatives to purchase agricultural equipment, fertilizer, livestock, and seed.[90] In 1959, some 189 farming cooperatives were in existence, and their membership corresponded with the social character of each colony.[91] For example, some colonies (and the relevant cooperative societies) comprised mostly displaced persons or political sufferers.[92] The UP government also organized a cooperative sugar factory in the Nainital tarai town of Bazpur and a consumers' cooperative store in Kashipur, which sold rationed supplies to its members.[93]

The Colonization Department created the Tarai Vikhas Sakakari Sangh ("Tarai Development Organization"), an organization whose purpose was to "assist the members of all these [189] societies in cultivation of the lands in all respects other than credit facilities."[94] Individual cooperatives were based on one colony: a village of roughly 100 families who occupied between 1,000 to 1,500 acres. The cooperative provided its members with the means of cultivation, including bullock plows and tractors. In addition, the cooperative provided for the harvesting and marketing of crops for the village.[95] One such agricultural cooperative was that of the Dhaulpur colony. This colony was populated by refugees from Punjab and received a range of assistance from the UP government. For example, the settlers received loans at 3.15 percent interest to build houses. In addition the cooperative received a Rs. 20,640 loan in 1952-1953 to purchase farm equipment.[96] As early as 1949, the Reserve Bank of India concluded that cooperatives were an important part of the colonization scheme in Nainital in which "[t]he settlers have braved the danger which pioneers in every field have to face and have converted lands once barren into smiling fields."[97] It was, however, premature to pronounce the tarai colonization project a success after one year of existence since most of the colonists involved had yet to take up their allotments, and the Tarai State Farm had just been established.

By the 1960s, many of the cooperatives in the tarai were no longer related to the colonization project with most of the cooperatives being created after 1961 for the purpose of avoiding the land reform laws. The members of these newer cooperatives were often joint families who sought to avoid losing land under the land ownership ceilings.[98] This constituted another deviation from the preferred landscape envisioned by Pant, Katju, and the TBDC where the members of the cooperatives were small-holding farmers who pooled their resources for the benefit of all.

Trees and Wild Animals

By the mid 1950s, the colonization of the Nainital tarai was well established: all of the colonists had taken up their allotments and the various cooperatives and the Tarai State Farm were fully in operation. As such, other issues in the colonization scheme arose, such as the afforestation or reforestation of parts of the colonized area and the problem posed to agriculture by wild animals.

As noted above, reclamation operations had been conducted on a large scale and thousands of acres of trees had been cleared. The Colonization Department, therefore, conducted afforestation programs to provide colonists with fuel and fodder reserves. The Department's goal was to provide each village with a forested area of 25 acres, and by April, 1953 the Colonization Department had afforested 3,037 acres.[99] The Colonization Department also planted trees along 112 miles of roads in the tarai.[100] Many of the fuel and fodder reserves were managed by the UP Forest Department rather than the settlers.[101]

Wild animals were considered to be a problem for the colonization scheme from the beginning. As land was reclaimed beginning in 1948, the wild animal population in the tarai declined as their habitats disappeared. Farmers also used a number of crop protection measures including the erection of barbed wire fencing and the use of firearms.[102] Wild animals, however, continued to pose problems for farmers after the initial stages of reclamation as elephants, deer, and wild pigs ate crops, and tigers and leopards attacked cattle.[103]

To solve the problem of wild animals attacking crops and livestock, the UP Minister of Agriculture, A.K. Sherwani, proposed in January, 1947 that hunting be permitted in Hailey National Park. Sherwani argued that the park was a sanctuary for crop-destroying pests. He further argued that the park was overcrowded and that was why the animals were leaving the park and attacking the farms in the tarai colonization area. He also stated that the animals of Hailey Park were being reserved to "provide sport for some big men in the province."[104] This proposal lead to a protracted debate among Sherwani and the Agriculture Department, the Forest Department, individual colonists, and the Kumaun Development Board. Devaki Nandan Pande, a member of the Kumaun Development Board, argued that "'Hailey Park' is proving a curse for the villages adjoining the park....The cultivators have to maintain a double watch during the day and night; but they still find themselves unable to protect their crops and cattle from the damages done to them by wild animals spreading out from Hailey Park."[105] W.T. Hall, the Chief Conservator of Forests, argued that Sherwani misunderstood the facts surrounding the park. He noted that hunting in the park was against the law even for "big men" and that the UP government was officially committed to wildlife preservation as it gave an annual grant of Rs. 1200 to the Association for Preservation of Game in UP.[106] He also argued that several measures were already being taken to protect farmers living near the park and noted that villagers were legally permitted to shoot any wild animal in their fields and that shooting in zamindari forests near the park had been increased.[107] Furthermore, D. Davis, the Conservator of Forests, Western Circle (which included Nainital district), argued that the wild animal population in the reserved forests and the park was declining and that complaints by villagers about wild animals was "merely a ruse" to hunt animals in the park and surrounding reserved forests since "dead animals (especially cheetals) fetch handsome prices these days."[108]

M.L. Malhotra, a Nainital tarai colonist, stated that despite all their efforts tarai colonists were unable to protect their crops from deer and wild pigs. He argued that the colonists would be better able to help India overcome its foodgrain shortage if more were done to deal with wild animals. He suggested that one way to reduce the threat of wild animals was to cut down the zamindari forests near Hailey Park, which served to harbor deer and pigs.[109]

The Forest Department mounted a robust defense of the park and nearby forests against the continued criticism of the Agriculture Department and individual settlers. Hall argued that

> No shooting is allowed in the National Park and it provides sport for nobody whatever his status. The whole point of the Act is that no shooting of any kind will be allowed in it so that the wild animals may move about free from fear of man. The object of creating the Park was the preservation of wild animal life for scientific interest and so that the people in the Province could take a cultural interest in the wild life of the forest where the animals are free from destruction....if any kind of shooting is to be permitted in the National Park, the National Parks Act, 1935 might as well be repealed. In my opinion, such action would be lamentable [and not] in the cultural interest of the Province.[110]

In addition, Hall, in reply to Malhotra, argued that the area of forested land in UP was already too small and that any further deforestation would be imprudent.[111] D.P. Joshi, Divisional Forest Officer, Ramnagar Forest Division, suggested that the accounts of animal depredation of crops were exaggerated. He noted that a rinderpest epidemic had drastically lowered the deer population in the park and argued that the existing rules governing hunting should be retained.[112]

In the end, G.B. Pant mediated the dispute and decided that Hailey should retain the status of a national park, but that farmers who lived near it should be given more firearm licenses.[113] This issue was a difficult one for Pant, because, while he was personally opposed to big game hunting, he had publicly supported the right of farmers to protect their crops from wild animals since 1919.[114] In March, 1947, Hukam Singh Visen, UP Minister of Revenue, in a statement before the Legislative Assembly, articulated the state's official policy:

> The policy of Government is to preserve wild life, which is a national asset of scientific and general interest, provided that animals which are a danger to life, or which are liable to damage property, are kept within reasonable limits. It is also the policy of Government to allow controlled shooting by sportsmen of animals and birds in forests other than Hailey National Park.[115]

Indeed, in 1952 the Governor of UP and his wife shot three tigers in Nainital district in the Himalayan foothills just north of the tarai colonization area.[116]

Wild animals continued to be an issue after the debate over Hailey National Park in 1947. For example, in May, 1951, the UP government, advised by an "expert" from Orissa, undertook a project to capture elephants which were responsible for "considerable damage" to crops in the Nainital tarai.[117] Elephants, however, continued to be a problem, so in April, 1952, the

UP government proposed the shooting of problem elephants who were damaging sugarcane fields in the tarai.[118] Critics noted that this proposal was out of step with the state's commitment to wildlife preservation so in August, 1952, the state decided to capture and domesticate troublesome elephants.[119] Tigers were also considered to be a problem in the tarai, and in April, 1953 villagers near Rudrapur clubbed to death a tiger they claimed to be a man-eater.[120] In August, 1954, the UP government took two steps to assist farmers in protecting their crops from wild animals. The first was the creation by the Agriculture Department of several "self-help" squads of farmers stationed in villages throughout the state, including the tarai, which were to deal with troublesome wild animals.[121] The second was an initiative to grant more firearm licenses to farmers coupled with a publicity campaign to educate farmers that nilgais (meaning "blue cow" in Hindi) were not real bovines and that shooting them was not a sin.[122]

Bureaucratic Complications in the Nainital Tarai Project

The colonization of the tarai was a collaboration between several UP government departments, i.e. the Colonization Department, the Agriculture Department, the Forest Department and the Public Health Department; the GOI; and various United Nations agencies, like the FAO and the WHO. As might be expected under such conditions, there was tension, even controversy, among the various departments and agencies. While the most notable contention was between the UP government and the CTO as they negotiated fees to be charged, deadlines to be met, and conditions to be observed, there was also tension among UP government departments.

As noted above, the UP government and the CTO signed a contract hiring the CTO to deforest and reclaim the land of the tarai. Very soon afterward, however, both sides sought to make adjustments in the contract. One issue was the insistence of the CTO in 1949 that the UP government provide its employees in Kashipur with exclusive access to medical facilities that were separate from those provided to settlers and UP government workers. The CTO argued that such facilities were required by the CTO-UP contract and that it would not begin work in Kashipur in 1949 until they were provided.[123] The UP government initially objected because it felt that adequate medical facilities were already present in Kashipur, but eventually agreed to provide the facilities due to concern that failure to do so would delay reclamation operation.[124]

Another issue which arose in 1950 had to do with who would pay for the CTO's overhead expenses. The UP-CTO contract had required that the CTO work on a "no-profit-no-loss" basis, but the UP government was alarmed by the sharp rise in CTO fees during 1950. The CTO's fee for "tractoring" had risen from Rs. 100 per acre to Rs. 130 per acre and the UP government wanted the CTO to differentiate between overhead and operational expenses before it would make any further commitments.[125] Furthermore, the UP government objected that the CTO rates it was being charged did not match the rates quoted by the GOI to the World Bank when the GOI received a loan to buy equipment.[126] The GOI replied that it had never quoted specific figures in its loan application and that it would pay for the CTO's expenses and then recover the cost from the state government, which in turn would recover the cost from the colonists.[127]

A third issue was that the CTO began to raise its fees again in 1951. The UP government was alarmed when the CTO informed it that its fee per acre would vary between Rs. 130 and Rs.260 depending upon the terrain of the land to be recovered. The UP government believed that the higher rates meant that it would be subsidizing the inefficiency of the CTO's operations in other states. It also stated that land reclamation costs were higher in states like Madhya Pradesh and that if the CTO charged each state an appropriate fee then the cost of the CTO operation in the tarai would decline.[128] A.N. Jha, the UP Food Production Commissioner, also argued that the state's budget simply could not cope with the 100 percent increase in CTO fees and that had it known that the estimate of Rs. 130 per acre was inadequate it would "have made alternative arrangements" for the reclamation of 31,000 acres in the tarai.[129] The UP government was also concerned that the higher rates would place an undue burden on settlers who would eventually have to pay the reclamation costs.[130] Eventually the UP government and the CTO agreed that the GOI would subsidize any costs that were higher than the agreed Rs. 130 per acre. The GOI Finance Ministry, however, decided to treat the subsidy as a loan that was to be repaid by the UP government (and hence the tarai settlers).[131]

The relationship between UP and the CTO changed significantly when a controversy erupted in the GOI Lok Sabha in May, 1954 regarding the loss of money by the CTO. The Lok Sabha's Estimates Committee charged that mismanagement of the CTO had lead to the improper and uneconomical purchase of equipment and that this had lead to the sharp increases in CTO fees.[132] In a statement delivered to the Lok Sabha on May 21, 1954, Rafi Ahmed Kidwai, GOI Minister of Food and Agriculture, said that the criticism of the CTO had been inaccurate. He argued that the CTO had not mismanaged the purchase of equipment. Kidwai admitted that the CTO had made some mistakes, but argued that such mistakes were inevitable in a newly created organization that lacked any precedent. He argued that it was necessary to differentiate between honest errors in judgment and improper behavior. He asked the house, "Is it not correct that the Central Tractor Organization does represent an enterprise of a magnitude never before undertaken in this country?" Kidwai then promised to investigate the situation and punish anyone guilty of "culpable negligence."[133] Historian S.P. Singh, however, has argued that the CTO suffered from significant and basic mismanagement in the purchase of equipment, as well as the setting of organizational policy. He contends that the CTO's policy of "no-profit-no-loss" was unworkable, but that the CTO did not realize this until it was "quite late."[134] Once the organization realized this, it lowered its fees to the states.[135] The CTO then began to lose money, but the GOI considered this to be a subsidy in connection with the Grow More Food campaign.[136] Singh concludes that "there is no doubt that bad management and absence of definite decisions were mainly responsible for the failures of the organization."[137] The GOI decided to close down the CTO in 1959 as it had outlived its utility.[138]

There was also tension in the UP government among the various departments involved with the tarai colonization project. One such controversy was between the Colonization Department and the Forest Department. Beginning in February, 1950, the Forest Department charged the Colonization Department with the widespread encroachment of its land.[139] The Forest Department's main concern was that "the total forest area under State control in the Gangetic plain is dangerously low and any further deforestation of forests would not be in the best interests of the State."[140] G.B. Pant ordered the Colonization Department to explain why it encroached on Forest Department land repeatedly and stated that Colonization Department "staff should strictly observe the rules otherwise action will have to be taken against them and they would not be treated as a privileged class."[141] Radha Kant replied that the encroachments were due to honest error: the revenue records contained false information, and the terrain of the area made it difficult to recognize the boundary between Colonization Department land and the reserved forests. He also promised that it would not happen again and noted that the Colonization Department was committed to increasing the forested area of the state, since it had budgeted money for afforestation in the tarai.[142] The matter was settled when the encroached land was transferred to the Colonization Department.[143]

The third complication in the reclamation of the Nainital tarai was the problem of cost overruns. Due to the changes made in the tarai reclamation project after the publication of the TBDC report, the UP government faced a cost overrun of Rs. 99.99 lakhs by 1953.[144] A related issue was the question of collecting rent from the colonists. In 1951, the Colonization Department wanted to raise the rent of the settlers following the recent increase in the price of grain. Radha Kant argued that the UP government had spent a great deal of money to reclaim the land so that "[i]n such exceptional circumstances, there should be no objection to the modifying of the...rates with a view to bring up the rents to a proper level. Other zamindars cannot ask for the raising of the settlement rates because they have done nothing to improve the area anywhere on such a vast scale as this."[145] The issue was not immediately resolved because of the chaotic nature of the land and revenue records. One example of this chaos was that during reclamation operations, the village boundary pillars had been "ploughed up" and the Colonization Department concluded that it was impossible to fix village boundaries.[146] The UP Board of Revenue, therefore, would not entertain the question of changing the rents of settlers until a complete resurvey of the area had been made. In fact, after the resurvey had been completed, the UP Council of Ministers began to examine a reduction of the rent owed by settlers. Some ministers argued that the initial rent of Rs. 6/- per acre had been fixed in an ad hoc manner and the complaints of the settlers about it were justified. The settlers argued that not only did they have to pay rent for their land, but they also had to pay fees for irrigation water.[147] In the debate, the Finance Department opposed the reduction but the Revenue Department supported it. The Revenue Department argued that the settlers were already heavily indebted, that the collection staff had great difficulty in collecting the money from the settlers, and that it feared another depopulation of the tarai if the ad hoc rents were not adjusted.[148] In 1957, a decrease in rent was approved by the UP government.[149]

Unable to increase the rent of the settlers, the UP government sought to decrease the cost of the reclamation project. G.B. Pant appointed a committee in December, 1950 to investigate possible ways to reduce the expense of colonization. The committee presented its findings in February, 1952 and concluded that to save money the Colonization Department should reduce its staff and that the cost of settlers' houses should be reduced from Rs. 3500 to Rs. 1100. The committee also noted that the State Tractor Organization needed to become more efficient. These recommendations were accepted by the government, and some of them were implemented before the publication of the report in 1952. For example, the reduction of Colonization Department staff in the Nainital tarai was completed in 1950.[150]

A fourth complication was the issue of the allotment of land made to ineligible persons. An example was the land allotment made to Septa Farms

Cooperative. The UP government had given the cooperative an allotment of land near the tarai town of Gadarpur, but charged the cooperative with several violations of regulations in 1953. The UP government argued that the cooperative claimed to own the land it occupied even though it merely leased it. Furthermore, the government claimed that the cooperative had both encroached upon land held by Buxas and failed to pay rent and land reclamation fees. The cooperative, whose members were college graduates, denied the charges, but lost the land when the government refused to renew the cooperative's lease after the cooperative failed to pay some Rs. 51,000 in back taxes and reclamation fees. Upon a review of the situation, the UP government came to the conclusion that the cooperative's members received the land allotment improperly due to the support from "illustrious persons" associated with the GOI.[151] The initial supporters of Septa Farms included Datar Singh, Vice-Chairman, Indian Council of Agricultural Research and Additional Secretary, GOI Ministry of Agriculture, and V.V. Giri, GOI Minister of Labor, and later President of India.[152]

Colonization of the Tarai in Kheri and Pilibhit Districts

The reclamation of the tarai in Kheri and Pilibhit involved a much smaller effort on the part of the UP government than in Nainital district. Even though the government was instrumental in malaria control and some land reclamation, independent settlers played a larger role in the transformation of the tarai in Kheri and Pilibhit than in Nainital.

Due to the vast expense involved with the reclamation of the Nainital tarai, the UP government delayed reclamation work in Pilibhit and Kheri until the mid 1950s when it received financial assistance from the GOI to begin reclamation work in Pilibhit and Kheri.[153] The UP government's reclamation tarai land in Pilibhit and Kheri began with the passage of the Land Utilization Act in 1952. The UP government required land owners in Pilibhit and Kheri to cultivate the wastelands in their possession or have them confiscated and reclaimed by the state.[154] At the same time the UP government provided farmers in Pilibhit with loans to purchase tractors.[155] Irrigation was available from private tubewells and the Sarda Canal.[156] In 1955, the UP government began efforts to reclaim land in Nighasan tahsil, Kheri district, and eventually 7,500 acres were reclaimed. Six hundred families of landless agricultural laborers were allotted plots of 10 acres each, and 75 families of the "educated unemployed" were given plots of 20 acres each.[157] Also in Kheri district, 70 families of "Ex-criminal tribes" received 580 acres of land near Bhatpurwa village.[158] In Pilibhit a land reclamation project was started in 1958 and reclaimed 10,000 acres and involved 161 families.[159] In a land use survey published in 1962, however, the GOI argued that while there had been progress in land reclamation, there was nearly 8,500 acres of tarai land still available for reclamation in Pilibhit district.[160] These land reclamation projects in Kheri and Pilibhit were completed in the early 1960s.[161]

As in Nainital, malaria control efforts were a central part of the reclamation of land in the tarai of Kheri and Pilibhit districts. Malaria control operations began in Pilibhit and Kheri districts in 1952.[162] One difference between malaria control in Nainital district and Kheri and Pilibhit districts was the heavy reliance on voluntary labor in the latter.[163] These operations received financial and material assistance from the WHO, UNICEF, and the US Technical Cooperation program, but were separate from the National Malaria Control Programme, which was not introduced into Kheri and Pilibhit until 1954.[164]

The End of Agricultural Colonization in the UP Tarai

With the end of the state's second Five Year Plan in 1961, the government of UP ceased to emphasize the reclamation of wastelands as a way to increase agricultural production. As the UP government wound down the reclamation effort it took a number of steps to regularize the administration of the tarai. The first step was the removal of the Nainital tarai from the jurisdiction of the Colonization Department in 1953. The tarai of Nainital was placed under the regular governance of the Nainital district administration.[165] In 1954, the post of Director of Colonization was terminated, the Deputy Director of Colonization was named the Deputy Director of the Tarai State Farm, the Tarai State Farm was given responsibility for land reclamation in the tarai, and the Revenue Department was given jurisdiction over the work previously done by the Food Production Department and the Colonization Department.[166] Also in 1954, the UP government closed the STO amid allegations of improper spending.[167]

In the late 1950s steps were taken to close the Tarai State Farm and construct a land grant agricultural university on the site. To begin this process in 1958, the UP government passed The Uttar Pradesh Agricultural University Act, 1958. The university has also served as the center for the Tarai Development Corporation, a farmers' production and marketing cooperative.[168]

Even by the end of the 1950s, the preferred landscape of Pant, Katju, and the TBDC and the physical landscape it inspired was on the wane. The major manifestation of this process was the rise of big farmers and their mechanized farms.[169] While grants of more than 10 acres were made to farmers who possessed tractors were made from the beginning of land reclamation in the tarai, this was not part of the original plan outlined in the TBDC report. The TBDC had envisioned a landscape where small farmers, who occupied 10 acre farms and were members of cooperatives, were the intended colonists. Instead, both large and small grants of land had been made and the recipients of these different allotments had developed separate political interests. For example, when the UP government proposed to enforce land ownership ceilings in the tarai (as part of the zamindari abolition law), the big farmers protested.[170] The big farmers argued that their mechanized farms would become uneconomical if the land ceiling was applied to them. In the end, the big farmers' protests were effective, and they did not lose a significant amount of land.[171] Another aspect of the divergence in political interests along class lines among colonists in the tarai was that the big farmers eventually came to be the employers of a large number of landless laborers from UP and Bihar.[172] The preferred landscape advanced by Pant and Katju gave way to the preferred landscapes of others, especially as the "green revolution" arrived in the tarai.

Notes

[1] Bandopadhyay 1997:161.
[2] Pant was so interested in the reclamation of the tarai that he traveled to the Nainital tarai town of Kichha on January 4, 1948 to inspect the tarai colonization project. *National Herald*, "Pant for Kicha [sic]" January 4, 1948, p. 3.
[3] FD file number 203/1948, p. 1. In October, 1949, Pant decided to continue this arrangement. RCD file number 775/49, volume 3, p. 3.
[4] RCD file number 270/49, volume 1, p. 1.
[5] RCD file number 270/49, volume 1, p. 4.
[6] Pandey 1982:135-137.
[7] GOUP, Bureau of Agricultural Information 1953:40.
[8] GOUP, Bureau of Agricultural Information 1953:41.
[9] GOUP, Bureau of Agricultural Information 1953:41.
[10] GOUP, Bureau of Agricultural Information 1953:41-42.
[11] GOUP, Bureau of Agricultural Information 1953:42.
[12] *The Pioneer*, "18 Lakh Cost to Rehabilitate 30,000 Refugees" November 30, 1947, p. 3 states that the UP government intended to reclaim 90,000 acres in the Nainital tarai.
[13] FD file number 203/1948, p. 9. See also Government of India Information Services 1950:3.
[14] FD file number 203/1948, p. 11.
[15] The non-commercially valuable trees were used for firewood. The Forest Department, however, had a difficult time disposing of the wood because of its poor quality and had trouble giving it away. ACD file number 352/50, volume 1, p. 73 and volume 3, p. 74.
[16] GOUP, Bureau of Agricultural Information 1953:53 and Randhawa 1980-1986, 4:59.
[17] ACD file number 352/50, volume 1, pp. 55-56.
[18] Randhawa 1980-1986,4:59.

[19] RCD file number 89/1949, volume 4, p. 1.
[20] GOI, Central Tractor Organization 1953:v.
[21] GOI, Central Tractor Organization 1953:2-4. According to GOI, Central Tractor Organization 1953:8, 99 percent of the reclaimed grassland was under cultivation in 1953. According to Srivastava and Chakrabarti 1952:382 the total area under cultivation in the Tarai Colonization Scheme area (in Nainital) rose from 9,703 acres in 1947 to 26,073 acres in 1951.
[22] GOI, Central Tractor Organization 1953:22.
[23] ACD file number 352/50, volume 1, p. 55 and volume 3, p. 74.
[24] ACD file number 352/50, volume 3, p. 91.
[25] Srivastava 1947:363.
[26] Srivastava 1950:163.
[27] Srivastava 1950:155.
[28] Chand 1961:83.
[29] Srivastava 1950:156.
[30] Srivastava 1950:157.
[31] Srivastava 1950:158.
[32] Srivastava 1950:162.
[33] Srivastava 1950:163.
[34] Srivastava 1950:162.
[35] Srivastava and Chakrabarti 1952:386.
[36] Srivastava and Chakrabarti 1952:388.
[37] Srivastava and Chakrabarti 1952:392.
[38] Government of India Information Services 1950:3.
[39] Randhawa 1980-1986,4:59.
[40] See GOI, Ministry of Health 1952-1953 annual report and RCD file number 775/49, volume 3, p. 61.
[41] Chand 1961:84.
[42] Issaris, Rastogi and Ramakrishna 1953:315.
[43] *The Indian Medical Gazette* 1951:27.
[44] *The Indian Medical Gazette* 1951:27.
[45] Chand 1961:83.
[46] Rahman, Singh and Pakrasi 1956:158-159.
[47] Rahman, Singh and Pakrasi 1956:161.
[48] Rahman, Singh and Pakrasi 1956:162.
[49] Chakrabarti 1955:58 and Rahman, Singh and Pakrasi 1956:162.
[50] Chakrabarti 1955:58.
[51] GOUP, Bureau of Agricultural Information 1950:15.
[52] GOI, Ministry of Information and Broadcasting, Directorate of Advertising and Visual Publicity (n.d.):4.
[53] Government of Madras 1959:124.
[54] See Chakrabarti 1955, Srivastava and Chakrabarti 1952, Issaris, Rastogi and Ramakrishna 1953 and Rahman, Singh and Pakrasi 1956.
[55] Chand 1961:87.
[56] RCD file number 775/49, volume 3, pp. 61-62.
[57] GOI, MII Annual Report for the Years 1956 and 1957, pp. 38-39. See also Livadas 1957:25 and Rao 1955:88.

[58] See RAD file number 387/1949, volume 1, pp. 14-15 for the case of one Lt.-Col. Sarkaria, who received such a special allotment. In 1950, G.B. Pant argued that "we will not be able to render any financial assistance or to advance any money to their (unregistered displaced persons) rehabilitation. We have not been able to make provision for many of those displaced persons who are registered and are facing difficulties in our province, and they have a prior claim on us." RAD file number 387/1949, volume 1, p. 42.
[59] RAD file number 387/1949, volume 1, p. 2.
[60] RAD file number 387/1949, volume 1, pp. 11, 14.
[61] RAD file number 387/1949, volume 1, p. 6.
[62] RAD file number 387/1949, p. 11.
[63] RAD file number 387/1949, volume 1, p. 11.
[64] Tarai Farmers' Federation (n.d.):iii.
[65] See Agarwal 1994.
[66] RAD file number 387/1949, volume 1, p. 11.
[67] RAD file number 387/1949, volume 1, p. 41.
[68] Randhawa 1980-1986,4:57.
[69] Randhawa 1980-1986, 4:58.
[70] CD file number 961/51, p. 1 and *The Pioneer*, "Rehabilitation of E. Bengal DPs" April 11, 1952, p. 3.
[71] CD file number 961/51, p. 3.
[72] *The Pioneer*, "East Pakistan DP's Welfare" June 6, 1952, p. 3 and *The Pioneer*, "Rehabilitation of East Bengal Refugees in Tarai" May 24, 1954, p. 3.
[73] RAD file number 387/1949, volume 1, p. 43.
[74] RAD file number 387/1949, volume 2, p. 28. "Political sufferers" were those members of the nationalist movement (including Congress activists) who had been jailed or otherwise punished by the British for nationalist activities. Generally, Radha Kant and other UP government officials used the term to refer to rank-and-file nationalists, not Congress leaders like Mohandas Gandhi, Jawaharlal Nehru, or Pant.
[75] FPB file number 331/1951, p. 1.
[76] FPB file number 331/1951, pp. 1-6.
[77] GOUP, Bureau of Agricultural Information 1953:92, 94.
[78] GOUP, Bureau of Agricultural Information 1953:95 and GOUP 1950:9.
[79] GOUP, Bureau of Agricultural Information 1953:95 and Randhawa 1980-1986,4:58.
[80] TD file number 50T(7)/1952, volume 1, p. 20.
[81] GOI, Central Tractor Organization 1953:89.
[82] *The Pioneer*, "Administration of Govt. Estates" June 3, 1952, p. 3. According to H.L.V. "Terai Bhabar and Ganga Khadir" *The Pioneer*, January 24, 1953, p. 4, 50,000 acres had been reclaimed by the CTO. 20,000 acres were allotted to settlers, 17,000 given over the Tarai State Farm, and 8,000 acres were set aside for afforestation.
[83] Randhawa 1980-1986,4:58.
[84] Randhawa 1980-1986,4:57 and GOUP 1957:6.
[85] Randhawa 1980-1986,4:57, GOUP 1957:6, GOUP 1950:9.
[86] GOUP, Planning Department 1962:39.
[87] ACD file number 352/50, volume 2, p. 48.
[88] GOUP 1957:6 and GOUP, Planning Department 1962:39. A maund equals 40 seers or 82.28 pounds or 37.35 kilograms. According to GOI, CTO 1953:88 the Tarai State Farm generated a profit of Rs. 8 lakhs in 1950-1951 and Rs. 12 lakhs in 1951-1952.

[89] GOI, Reserve Bank of India 1949:11 and Government of Madras 1959:118.
[90] RAD file number 387/1949, volume 1, p. 11, GOUP 1953:16 and GOUP, Bureau of Agricultural Information 1953:98.
[91] Government of Madras 1959:118.
[92] ACD file number 352/50, volume 4, p. 183.
[93] GOUP, Bureau of Agricultural Information 1957:18 and GOUP, Registrar of Cooperative Societies 1954:21.
[94] Government of Madras 1959:124.
[95] GOUP, Registrar of Cooperative Societies 1954:19-20.
[96] Government of Madras 1959:118-119.
[97] Government of Madras 1959:51.
[98] Jalal, Bisht and Elhance 1984:227-228.
[99] UP Bureau of Agricultural Information 1953:15 and 134 and FD file number 203/1948, p. 69. Also in 1953, Radha Kant ordered that "[n]o area of the forest land is to be deforested now." FD file number 203/1948, p. 69.
[100] S.S. Negi "Afforestation Outside the Reserved Forests" *The Pioneer*, December 12, 1954, p. iii. Negi was the Conservator of Forests, UP.
[101] FD file number 574/51, p. 32.
[102] Firearms were also issued to Colonization Department workers during deforestation and reclamation operations. FD file number 203/1948, p.2.
[103] UP Bureau of Agricultural Information 1953:36-37 and GOI Information Services 1950:3.
[104] FD file number 189/1947, volume 1, pp. 1 and 5.
[105] FD file number 189/1947, volume 2, p. 32.
[106] FD file number 189/1947, volume 1, p. 7.
[107] FD file number 189/1947, volume 1, p. 34.
[108] FD file number 189/1947, volume 1, p. 42.
[109] FD file number 189/1947, volume 2, p. 45.
[110] FD file number 189/1947, volume 2, p. 2. It should be noted that fishing was permitted under certain conditions in the park. See E.P. Gee "The Tigers of Hailey National Park" *The Statesman*, October 10, 1954, Sunday Magazine, p. 1.
[111] FD file number 189/1947, volume 2, p. 51.
[112] FD file number 189/1947, volume 2, p. 45.
[113] FD file number 189/1947, volume 1, p. 15.
[114] Pant's objection to big game hunting was so well known that it was the subject of an editorial cartoon in *The Pioneer*, October 7, 1954, p. 4. Pant was also against fishing. In a letter to his children, dated March 17, 1943, he inquired as to the condition of Nainital town's lake and its fish. He wrote, "Similarly I fancy that the fish have not lost their charm and still swim and frolic with equal vigour. It is a pity that hard-hearted anglers will not allow them to live in peace....Fish are perfectly harmless and do none any wrong—just cruel people make fun of killing them, and do not feel any scruple even in devouring them! Such brutalities are a blot on the human race. Let us hope that man will learn to behave better..." National Archive of India, G.B. Pant papers, accession number 1600, file number 2. In August, 1954, the UP government proposed the establishment of 27 wildlife sanctuaries and named one after G.B. Pant. According to *The Pioneer*, "Pant Sanctuary in Tehri" August 6, 1954, p. 3, this act "symbolises the deep interest the Chief Minister has always been taking in the development of the State forests." Pant also served as UP Forest Minister from 1952-1954. UP Forest Department 1961:11.

[115] FD file number 189/1947, volume 3, p. 7. It should be noted that the UP state government as part of the second Five Year Plan allotted money for the "extensive development of Corbett National Park..." UP Planning Department 1962:43.
[116] *The Pioneer*, "Governor and Lady Mody Bag Three Tigers" May 10, 1952, p. 4. The article noted that "The party had an exciting day. One of the three tigers was shot by Mrs. Mody."
[117] *The Pioneer*, "Capture of Elephants in UP" November 23, 1951, p. 3.
[118] *The Pioneer*, "Destruction of Wild Elephants in UP" April 2, 1952, p. 4.
[119] *The Pioneer*, "The Elephant Tractors" August 24, 1952, p. 3 and *The Statesman*, "Menace of Wild Elephants in UP" June 18, 1954, p. 3. The wild elephant population in Nainital district was estimated to be 300.
[120] *The Pioneer*, "Man-Eater of Kumaon Beaten to Death" April 12, 1953, p. 1.
[121] *The Pioneer*, "Protection of Crops" August 6, 1954, p. 3.
[122] *The Pioneer*, "Saving Crops from Nil Gais" August 6, 1954, p. 3.
[123] RAD file number 363/1949, volume 1, pp. 9-10.
[124] RAD file number 363/1949, volume 1, pp. 7-10.
[125] ACD file number 352/50, volume 1, p. 8. "Tractoring" included a variety of activities, including deep soil plowing, harrowing, and tree root cutting.
[126] ACD file number 352/50, volume 1, pp. 46-47.
[127] ACD file number 352/50, volume 1, p. 47.
[128] ACD file number 352/50, volume 1, p. 87.
[129] ACD file number 352/50, volume 1, p. 61.
[130] ACD file number 352/50, volume 1, p. 78.
[131] ACD file number 352/50, volume 1, p. 78.
[132] *The Pioneer*, "Tractor Organisation's Imprudent Policy" May 13, 1954, p. 4. There was an editorial cartoon in *The Pioneer*, May 14, 1954, p. 4 in which a poor-looking man labeled "taxpayer" delivered a powerful kick to the posterior of a grossly fat man labeled "CTO."
[133] *The Pioneer*, "CTO Muddle: Text of Statement" May 22, 1954, p. 6.
[134] Singh 1973c:277.
[135] Singh 1973c:277 and *The Statesman*, "Tractor Rates Reduced" August 30, 1954, p. 1.
[136] *The Statesman*, "Tractor Rates Reduced" August 30, 1954, p. 1.
[137] Singh 1973c:278.
[138] Singh 1973c:320.
[139] FD file number 203/1948, p. 40.
[140] RAD file number 363/1949, volume 1, p. 22.
[141] RAD file number 363/1949, volume 1, pp. 24-25.
[142] RAD file number 363/1949, volume 1, pp. 30-31.
[143] RAD file number 363/1949, volume 1, p. 33.
[144] GOUP 1953:15.
[145] RCD file number 54/1950, volume one, p. 18.
[146] RCD file number 54/1950, volume four, p. 16.
[147] RCD file number 54/1950, volume four, p. 75.
[148] RCD file number 54/1950, volume four, p. 78.
[149] RCD file number 54/1950, volume four, p. 84.
[150] CD file number 540/50, pp. 19-21.
[151] FPB 759/1951, volume 1, p. 108.
[152] FPB 759/1951, volume 1, p. 62.
[153] ACD file number 352/50, volume 4, p. 196 and GOI, Planning Commission 1957:329-330.

[154] *The Pioneer*, "Cultivable Wasteland May Be Seized" October 12, 1952, p. 1.
[155] *Indian Farm Mechanization* 1952:5.
[156] Singh and Misra 1965:31.
[157] GOUP, Planning Department 1959a:20.
[158] GOUP, 1957:20-21.
[159] GOUP, Planning Department 1959a:20 and GOUP, Planning Department 1960:20.
[160] GOI, Wastelands Survey and Reclamation Committee 1962:16-20, 45.
[161] GOUP, Planning Department 1962:40.
[162] GOUP, Information Directorate (n.d.):26, UP Medical and Public Health Department, Annual Report for the year 1952-52, p. 9 and *The Pioneer*, "Anti-Malaria Drive in Pilibhit" October 5, 1952, p. 2.
[163] On p. 18 of the Republic Day Supplement of *The Pioneer*, January 26, 1953, the UP government had an advertisement entitled "Employment of Voluntary Labor in Control of Malaria in UP." The government used volunteer labor in its malaria control projects in 24 districts, including Kheri and Pilibhit. The government described the use of volunteers as an "experiment."
[164] *The Pioneer*, "Anti-Malaria Drive to be Intensified" December 1, 1952, p. 3, *The Pioneer*, "Anti-Malaria Measures in Uttar Pradesh" February 19, 1953, p. 3, *The Pioneer*, "Malaria Control in Uttar Pradesh" June 11, 1954, p. 3, *The Pioneer*, "Anti-Malaria Operations in UP Districts" August 7, 1954, p. 3.
[165] RCD file number 775/49, volume 2, p. 89. See also FD file number 576/51, volume 2, p. 35.
[166] RCD file number 775/49, volume 3, p. 82.
[167] *The Pioneer*, "Tractor Organisation Being Reorganised" March 17, 1954, p. 5.
[168] University of Illinois at Urbana-Champaign, Department of Architecture 1975:11.
[169] The GOUP, Bureau of Agricultural Information 1950:50 describes big farmers as "capitalists who have taken out large leases in these areas [in the Nainital tarai]."
[170] See *The Pioneer*, "Land Reforms Soon in Govt. Estates" July 3, 1954, p. 3.
[171] See Tarai Farmer's Federation (n.d.), P.N. Mehta "Mechanized Farms" *The Pioneer*, June 5, 1954, p. 4, N. Prakash "Case for Tarai Farmer" *The Pioneer*, June 18, 1954, p. 4 and Bakshi Tirath Ram "Tarai Farmers' Prospects" *The Pioneer*, July 7, 1954, p. 4. According to an informant interviewed by the present author in New Delhi in 1994, the Tarai Farmers' Federation's membership was dominated by big farmers.
[172] Swarup 1991:43.

• CHAPTER SIX •

Outside the Official Paradigm

The government of Uttar Pradesh was not the sole agent in the transformation of either the tarai's preferred or physical landscapes. The process of reconstructing the landscape was complicated and influenced by the actions of influential independent settlers, landgrabbers, and unlicensed hunters or "poachers". These non-officials played a significant role in defining how the tarai would be integrated into the physical landscape of northern India. These groups developed their own interpretations of the tarai that varied, sometimes substantially, from the official vision as defined originally by G.B. Pant, K.N. Katju, and the Tarai and Bhabar Development Committee. Hence, the government of UP was not the only agent in the normalization of the region, yet non-officials were not necessarily at loggerheads with officials. Poachers, for example, influenced and were influenced by a shift in the preferred tarai landscape in which less and less value was placed on the conservation of wildlife. Poachers had a pronounced impact on the faunal life of the tarai, which was already decreasing as the result of the loss of habitat due to official land reclamation. Outside of Jim Corbett's books, the tarai was no longer home to vast wild animal populations. In a sense, poaching accelerated the process of normalization the tarai, because of its contribution in lowering the wild animal population (outside of the national parks) to levels generally found in the plains of northern Indian. The tarai, in large measure, ceased to be distinctive in the sense that it no longer possessed an almost endlessly large population of wild animals. Of course, wild animals continued to live in the national parks of the tarai, but, after national parks were established elsewhere throughout India, the tarai's wildlife ceased to act as a marker for the exotic or unusual.

Further, a significant amount of the land reclamation was conducted by private colonists independently of the official process. These independent colonists included retired army officers as well as Sikh and Hindu Punjabi refugees, some of who anticipated the partition of the Punjab and arrived in the tarai before August, 1947. Often these settlers obtained land from the local zamindars, but others became tenants of the Tarai and Bhabar Government Estates in 1945 and 1946. Still others occupied land illegally, either by encroaching on forest land or by seizing it from the Tharus and Buxas. This group of settlers, the landgrabbers, were eventually incorporated into the official colonization process regularizing their holdings or by government efforts to evict them. Independent settlers, including those who engaged in landgrabbing, had a distinct perception of the tarai, one that differed sharply in some ways from that of the UP government. For example, as noted earlier, when the UP government adopted the Tarai and Bhabar Development Committee's report, it wanted tarai colonists to be small-holding members of agricultural cooperatives. The independent settlers often had significantly larger holdings than those who had gained land through the official process, and many were not members of cooperatives.[1]

Independent Settlers in Kheri and Pilibhit

Perhaps the most notable private settler in the tarai was Arjan "Billy" Singh, scion of a Punjabi princely family and a retired army officer, who settled in Nighasan tahsil, Kheri district in 1945 after Indian army service in the middle east.[2] When he arrived in the area he observed that in the northern portions of the tahsil, agriculture was severely restricted due to the activities of dacoits who were based in the North Kheri Forest Division and even lived in Forest Department facilities in the Forest Division during the rainy season. With the added pressure of malaria on top of dacoity, villagers had fled the area. In the southern portion of the tahsil, there was little cultivation, although there was a significant population of graziers. According to Singh, "[t]he sparse population of the district was now [1945] huddled in a few villages, racked by seasonal fevers, devoid of ambition and content simply to exist as their ancestors had before them."[3] It should be noted, however, that not all farmers in Nighasan lived in such wretched conditions. The family of journalist and conservationist Rahul Shukla began to profitably farm 100 acres near present day Kishanpur Wildlife Sanctuary sometime after 1900. Shukla observes that "[t]here was no agricultural development in this region and the wild animals roamed everywhere. Notorious gangs of dacoits lived in the vicinity. A multitude of poisonous snakes together with the dreaded diseases malaria, jaundice and typhoid regularly claimed a substantial toll of human lives."[4] Shukla, however, does not discuss the condition of the indigenous population or the arrival of Arjan Singh to Nighasan tahsil.

When he arrived in Nighasan tahsil, Singh was aware of the tarai's reputation for dacoits and malaria but was not deterred.[5] Singh constructed a preferred landscape in which he believed that he could transform the wilderness into farmland. Key elements in his preferred landscape are found in his observation that "[t]here was any amount of land available and the shooting was out of this world."[6] He felt that the tarai could provide him with opportunities for agriculture as well as shikar. Singh bought 750 acres near the town of Palia in southern Nighasan and he named his farm "Jasbirnagar." He found the use of animal-drawn plows very ineffective in breaking the soil, with its long grass, so in 1946 he purchased a tractor. With the help of Micky Nethersole, Senior Member, Board of Revenue, Singh received a government loan to buy the tractor, a John Deere Model B.[7] While Singh was able to secure a government loan, other settlers in Kheri district were not. In 1948, the UP Relief and Rehabilitation Department decided to allot funds to certain districts for advancement of loans to refugees by district magistrates. This money was available to settlers in Pilibhit and Nainital districts, but not in Kheri district.[8]

As he sought to alter the physical landscape through the cultivation of rice and sugarcane, Singh faced two major competitors: the indigenous Gaddi population and wild animals. That is to say that while Singh sought to realize his imagined tarai, he faced the challenge of the Gaddis who sought to realize their own interpretation of the landscape. The Gaddis, who were Muslim graziers, and Singh engaged in a rivalry over the area's resources. The Gaddis were accustomed to grazing their cattle freely and resented the presence of Singh. They interpreted the agricultural prospects of the area solely in terms of animal husbandry and resisted Singh's attempt to shift the physical landscape in the direction of cultivation. The Gaddis and Singh were soon engaged in litigation, but, according to Singh, "by a mixture of cajolery and strong-arm tactics I held my own, and gradually the cattle owners realized that agriculture had come to stay, and soon the Gaddis themselves turned their efforts to farming."[9] Singh, however, has not discussed any Gaddi interpretation of these events.

In 1945, the tarai area outside the North Kheri Forest Division teemed with wild animals, such as foxes, jackals, tigers, fishing cats, leopards, wolves, hogdeer, nilgai, black buck, elephants, crocodiles, snakes, and pigs; the introduction of cultivation to the area did not come easily.[10] Singh had great difficulty with various herbivores attacking his crops and he shot numerous animals, especially pigs, in order to protect crops, but he was able to drive off elephants and deer with loud noises.[11] Rahul Shukla's family also faced the threat of wild animals. He describes the scene:

> The wild pigs were an especially incorrigible nuisance. They dug up potatoes and peanuts and often damaged a greater quantity than what they could actually consume. The pragmatic solution to this problem lay in the elimination of these raiders. The farmers were troubled, for the local supply of wildlife was so great that bags of

bullets were emptied upon them, yet the animals continued to pour in. Hunting also continued as an effective remedy to the menace.[12]

The wild animal population in southern Nighasan slowly declined through such crop protection measures by the new colonists and cultivators as well as through the loss of habitat signified by the transformation of grassland into farmland.

By 1950, other settlers, mostly Punjabis, arrived in the southern portions of Nighasan tahsil. The official colonization process did not begin in this area until 1952, with the start of anti-malaria operations, so these independent colonists were able to lease land from the local zamindars without restrictions on the size of the holdings. Initially, the Punjabi colonists suffered heavily from malaria, but they persevered. Singh relates that while whole families died from malaria, others stayed on "because of the promise of large holdings and the prospect of converting the waving seas of grasslands into productive crops."[13] In the preferred landscape constructed by these settlers, the tarai had the potential to be converted from wasteland to productive farmland through their own labor. In other words, they possessed their own interpretation of the tarai and sought to implement it by conducting reclamation activities without state assistance.

According to Singh, settlers concentrated in southern Nighasan and avoided the northern portions of the tahsil due to the virulent form of malaria endemic there.[14] As they moved into the southern part, the wildlife shifted northward towards the North Kheri Forest Division, marking a significant change in the physical tarai landscape. According to Singh, "[t]he herds of black buck and nilgai retreated northwards before the advance of crop-protection firearms, and with them went the big cats which preyed on deer."[15] Singh has variously described the transformation of the area:

> When I first came to Jasbirnagar my farm was an island surrounded by an ocean of waving, ten-foot grass; but gradually, with the influx of sturdy husbandmen from the Punjab—from which they had been displaced by the holocaust after Partition—my land became merely one piece in the jigsaw of the agricultural landscape.[16]

> [By the late 1950s] the landscape was changing. Whereas I had come to an endless vista of grassland, multi-coloured agricultural crops now shimmered on the horizon. Gone were the herds of Nilgai and blackbuck, and somehow the midnight cries of the fox and jackal seemed muted.[17]

Rahul Shukla also describes this process: "The gnawing away of the habitat had consequently affected the plethora of avian and ungulate life."[18] Shukla observed that the grasslands were replaced by maize fields and that migratory birds ceased to stop at local marshes and streams as "hordes of Sikh refugees" arrived in the area after independence.[19] According to Shukla, "[t]he great forests and grasslands of Kheri which for centuries had been the traditional breeding grounds of tigers succumbed to the ploughers and harvesters."[20]

Independent settlers also colonized land in Puranpur and northern Pilibhit tahsils, Pilibhit district in the late 1940s. The pattern of settlement in Puranpur and Pilibhit tahsils was similar to that in Nighasan, except that colonists in Puranpur and Pilibhit had access to government loans. By 1951, these colonists had brought some 1500 acres into production.[21] In 1951, the independent settlers around the town of Neoria in Puranpur tahsil petitioned the UP government to construct irrigation facilities in the area. They argued that they would be better able to participate in the state's Grow More Food campaign if they had access to irrigation. The Irrigation Branch of the Public Works Department flatly denied the request, stating that the addition of irrigation works would lead to water-logging of the soil due to the area's high water table which in turn would lead to the creation of addition mosquito-breeding grounds.[22] Other colonists in the tarai of Pilibhit district did eventually gain access to irrigation, mostly through private efforts. The village of Amaria, northern Pilibhit tahsil, possessed some 614 acres of irrigated farmland by 1971.[23]

"Crop Protection" and Poaching

Arjan Singh has argued that much of the "crop-protection" activities by the early settlers was actually poaching, especially as the wildlife began to retreat northward.[24] The methods of the poachers varied but in some cases they used machine guns mounted on jeeps to track and kill game.[25] In addition, Singh has charged that timber contractors, UP state officials, and local police officers engaged in these poaching activities.[26] Indeed, these groups had greater access to modern firearms than the Gaddis, who possessed only muzzle loading rifles.[27] Due to the differing levels of firepower, the poaching by police and other governmental officials had a greater impact on the wild animal population, and thus the physical and preferred landscapes, than poaching by the Gaddis. However, the GOI has blamed "tribals" for much of the poaching. Foreign tourists who came to India prior to 1972 to shoot tigers were seen as "legitimate" hunters, while adivasis who engaged in hunting for subsistence needs were condemned as "poachers."

Conservationist Rahul Shukla agrees with Singh that while early settlers did have genuine problems with wildlife, these "crop protection" activities had quickly become an excuse for poaching. He observes that by the early

1960s, "I found that the pretext of crop protection was not valid for this thoughtless, illegal execution."[28] He noted that items like "hog deer steaks," "rabbit soup," "peahen kababs," and "spotted deer fillets" were a routine feature of the menu of villagers in Nighasan tahsil.[29]

Poaching was not, however, confined to the tarai region, but was widespread throughout India, and the country's wild animal population dropped substantially after independence. P.D. Stracey, an Indian Forest Service officer, observed in 1957 that "[o]ne of the more striking facts of the last decade or more has been the rapid disappearance of wild life throughout India."[30] In 1967, wildlife researcher J.J. Spillett concluded that

> [p]oaching, combined with habitat abuse or destruction, has presently attained such proportions in much of India that with many species of Indian wild life it is now more a matter of preservation rather than sustained yield management....Unless [the poacher's] depredations are soon brought under control, much of India's priceless and irreplaceable wildlife will be lost forever.[31]

Of course, poaching frequently occurred before 1947 but after independence many Indians viewed the game laws as outdated relics from the Raj.[32] Ornithologist Salim Ali argued that "[w]ith Independence in 1947, the loosening of firearms controls and laxity in the maintenance of law and order—when most of the conservation-conscious British forestry and administrative officials left the country—India's wildlife fell on evil days."[33] Stracey claims that "[t]he emergence of a new class of shikaris with the free issue of licenses for weapons in the new flush of independence has brought a new [low] standard of sportsmanship."[34] Licensed hunters and poachers alike ignored all hunting rules and regulations, and in several states the enforcement of game laws was a low priority.[35]

In 1970 an expert committee convened by the Indian Board for Wildlife issued a report entitled, *Wildlife Conservation in India,* in which it denounced poachers and assigned to them a great deal of the blame for the decline in wild animal population:

> Animals have been trapped, netted, poisoned, shot and otherwise killed for their meat, ivory, horn, fur, and this illegal traffic, and occasionally even licensed hunting encouraged by the middleman who continues to make a large profit, has been responsible for so many species being endangered today. The use of motor vehicles with spotlights, inside and outside forests, takes a heavy toll of all wildlife. The village shikaris who prowl round with packs of dogs, and sit up over waterholes in summer, kill every animal and bird without the least concern for the law or for the future of the species.[36]

Furthermore, the committee concluded that "[t]he wildlife of India is approaching extinction" and that "[c]rop protection and self defence were common excuses used for widespread massacre, and unrealistic shooting rules and ineffective conservation laws, did the rest."[37] The committee, in agreement with Arjan Singh, observed that many policemen, military personnel, and diplomats engaged in poaching.[38] In addition, many species like wild dogs, leopards, pigs, and jackals were classified as "vermin" and a reward was offered to anyone who killed one.[39] At this time, wild animals were widely interpreted throughout parts of Indian society as having little value while alive, but great value once dead.

Independent Settlers in Nainital

In the tarai of Nainital district, independent settlers obtained land from local zamindars as well as from the Tarai and Bhabar Government Estates. During 1946-47, the Estates received a number of applications for land from refugees from western Punjab, but these applications were initially declined due to Revenue Department policy.[40] This policy was designed to keep unauthorized colonists from gaining land in the tarai colonization area before the official colonist selection process began. The government was concerned that some would-be settlers had obtained "fictitious leases" and wanted to exploit the colonization scheme. Officials like Radha Kant expressed interest in protecting genuine tenants from the loss of their land.[41] The government, however, quickly revised the policy and allowed private settlers who had obtained land after 1946 to colonize land in areas in the tarai outside the official colonization program. Geographer Shailaja Pandey has observed that "[o]utside the area of the colonization project from 1946 to 1950 enterprising individuals, including many army officers, had with the consent of the Superintendent of the Bhabar and Tarai...reclaimed the waste land, and built up agricultural estates of several hundred hectare size."[42]

The establishment of these large estates marked a shift in the government's preferred tarai. Originally the government desired a tarai which was to be populated by small holders with holdings varying between 10 and 50 acres. R.S. Bisht, Superintendent of the TBGE, in the Annual Report of the Administration of the TBGE for 1946-47, however, noted that few of the Punjabis who were allotted land "have actually turned up."[43] This failure to turn up may, in part, have been due to the tarai's deadly reputation. Harpal Singh Sandhu, Deputy Director of Colonization, observed that the tarai's reputation scared off numerous would-be colonists in the first years of the official colonization scheme. According to Sandhu, "[t]he Tarai after all was a notoriously inhospitable area—the *Kala Pani* (black water)—and nobody wanted to risk founding a home in Tarai."[44] If refugees were wary of settling land already reclaimed by the state, then it is understandable that other settlers were reluctant to reclaim land themselves.

Despite qualms, some independent settlers were able to obtain land in areas included in the official reclamation scheme prior to the beginning of the project. In Kashipur tahsil in Nainital, these settlers were granted exemptions and were permitted to keep the land they had obtained from local zamindars. Radha Kant explained in 1950 why refugees and other settlers had come to Kashipur tahsil: "Partly as a result of the knowledge that Govt intended developing the North Kashipur area and partly as a result of the anxiety of displaced persons to obtain lands for cultivation, there has been great land grabbing activity in Kashipur during the last two years."[45] The government based the exemptions on the ability of the private settlers to introduce mechanized cultivation. Furthermore, these settlers were able to keep every acre that they cultivated, even though they possessed holdings larger than the standard 15-acre allotment made to official colonists. The reasoning for this was that a 15-acre farm was too small to support the cost of mechanized cultivation. Those farmers who owned tractors were allowed to have at least 30 acres, the minimum seen by the government as necessary to support tractor farming.[46] One example was the exemption made for Major M.L. Malhotra, a disabled veteran from the Punjab. Malhotra obtained roughly 1452 acres in Kashipur tahsil in early 1949 and petitioned the Colonization Department for an exemption. Upon the recommendation of Kant, the government granted the exemption in June, 1950.[47] Again, the development of these mechanized farms represented a modification of the government's preferred tarai. These settlers were able to insert themselves into the Tarai Colonisation Scheme even though they had originally been excluded.

In Kashipur tahsil tension developed between the existing residents of Kashipur town and the Punjabi settlers in February, 1951. The residents of Kashipur complained that the Punjabi farmers had settled on land that they had traditionally used to graze their cattle. They further claimed that the Punjabis had molested their cattle. The Punjabis, however, had legally occupied the land. According to Radha Kant, "[s]o far as these Punjabi farmers are concerned, they took their leases from Zamindars and we simply recognized them."[48] Initially, Kant and other Colonization Department officials wanted to require the townspeople to either buy fodder from nearby farms or drive their cattle about three or four miles down the Kashipur-Bazpur road to an area by the Kosi river that had previously been set aside for grazing.[49] G.B. Pant overruled this decision and required that land closer to the town be found for the use of the townspeople "even at the cost of the adjustments in the colonization scheme."[50] In this clash of economic interests and associated preferred landscapes, Pant recognized that the government's original plans had to be modified again to accommodate the earlier decision to regularize the holdings of the Punjabi landgrabbers.

The new preferred landscape informed the alteration of the physical landscape as the Colonization Department then proceeded to survey the area around Kashipur town for uncultivated land that could be reserved for the use of the townspeople. The department identified some 1,388.69 acres, in three separate blocks, that it proposed to acquire and set aside for the purpose.[51] Some the occupants of this land did not wish to vacate their land and objected to the land acquisition proceedings. One Virinder Pal Singh, who had obtained his land legally, filed a case in the court of the Deputy Commissioner in Nainital town. Singh stated that he was a displaced person from Pakistan and had expended a great deal of effort and money on establishing his farm. He stated that he had no other land and would be financially ruined if his land was acquired by the government. He also pointed out that a great deal of other grazing lands were available to the residents of Kashipur town and that his land was not needed. In December, 1951 the government decided that it did not need to acquire all of the original 1,388.69 acres and scaled down its proposal. Singh was allowed to retain his land.[52]

When government land reclamation operations ceased by 1961, the amount of landgrabbing increased. This landgrabbing was done by immigrants to the tarai who were unable to participate in the colonization process. The land occupied by landgrabbers was sometimes forest land, including those areas set aside to serve as fuel and fodder reserves, but much of it was land seized from the Tharus and Buxas through various illicit means.

Amir Hasan describes this process of landgrabbing as the "willful dispossession of the Buxas of sizable portions of their land by hook or crook."[53] In other words, the adivasis lost out in the competition for resources. Hasan determined that by 1975 some 22 percent of Buxas in Nainital district had been affected by landgrabbing. Many of the Buxas were dispossessed of their land without any compensation at all, and most of them had lost their land under duress. In a 1968 report, the UP Tarai Lands Committee observed that "[l]and in Buxari villages has also been purchased irregularly or trespassed upon by more aggressive outsiders."[54] Landgrabbers used outright force, theft of crops and cattle, and indebtedness to part the Buxas from their land. According to Hasan, a majority of the landgrabbers were Punjabis, including Sikhs.[55] In Nainital district the Tharus and Buxas owned about 250,000 hectares of land in 1951; by 1993 they owned only about 30,000 hectares.⁵⁶ The Tarai Lands Committee estimated that some 20,000 trespassers grabbed around 53,000 acres of government land by 1968.⁵⁷ In 1978, the UP Social Welfare Directorate produced a report that claimed that some 1,300 trespassers had grabbed land from the adivasis.⁵⁸ In effect, landgrabbing was the large-scale reinterpretation of the tarai by certain immigrants to the tarai. Landgrabbers viewed forested land as wasted land and, thus, eligible for reclamation. This was in direct contrast to the government's interpretation of the tarai which held that forest land had a valuable place in the physical landscape.

The landgrabbing of adivasi holdings occurred elsewhere in the tarai. As noted above, the Gaddis of Kheri district lost access to grazing lands, for which they lacked legal title, but had customary or traditional access. The lack of legal title to land was a problem for adivasis and peasants throughout India, and the Gaddis, as an unscheduled tribe, had little legal recourse. In Nainital district, the Gujjars and other residents of the Khattas also held limited legal title to they land they occupied. Historian Ajay Rawat argues that several new residents of the Khattas, including retired army officers, government officials, local politicians, and other assorted "bigwigs," had grabbed forest land in the 1970s. These landgrabbers have vainly tried to get the UP government to regularize their actions in the same way that it had regularized earlier landgrabbing by settlers in Kashipur in the early 1950s.⁵⁹ In this case, the UP government was able adhere to its conception of the tarai, even though the physical landscape did not conform to the preferred landscape.

Reinterpretation of the tarai by other groups outside of the government continued as further landgrabbing was committed by landless laborers, dalits, and Bengali refugees. Journalist P.C. Joshi claimed that the "squatting on...uncleared fallow land" in 1969 in Nainital district by these landless poor people was "heroic" because of the violent hostility of Punjabi landgrabbers who had, allegedly, bribed local government officials to secure their own holdings.⁶⁰ These landless persons were organized by the Communist Party of India, and the hostilities between the two groups of landgrabbers became politicized. There have been violent clashes between the CPI-supported landgrabbers and the police. The People's Union for Democratic Rights, an NGO based in New Delhi, has alleged that the wealthy Punjabi landgrabbers have had close links with local police and politicians and have hired their own security guards.⁶¹ These Punjabis, both Joshi and the PUDR claim, prevented the dalits and refugees from taking up the lands they had been legally allotted. Arjan Singh, however, has a different interpretation and condemns the illegal occupation of land near the North Kheri Forest Division by a similar group of landless persons in 1966. He remarked in 1973 that these landless persons were Naxalites who "operate under the age-old principle of taking away from those who work and giving to those who talk..."[62] Even though these landgrabbers were later evicted from the land, Singh was unhappy that the UP government had not prevented them from occupying the land in the first place, because of the harm the landgrabbers caused to the habitat of the swampdeer, a threatened species.[63]

As noted above, land alienation was a problem for tribal populations throughout Uttar Pradesh. Adivasis (including scheduled and unscheduled tribes) and scheduled castes, and other forest-dependent populations lost land both to landgrabbing private citizens and to actions by the UP state government in acquiring land for various development plans. Adivasis have also suffered when frequently given inadequate compensation for land acquired by the government. According to Hasan, one serious problem for adivasis in UP is that most, like the Gujjars, are not formally recognized as scheduled tribes and do not receive the legal protections associated with scheduling. Additionally, thirteen tribes were erroneously listed as Scheduled Castes. The Bhotiyas, Tharus and Buxas, however, were not listed as Scheduled Tribes until 1967, which made it more difficult for them to deal with the landgrabbers. Since 1967 it has been illegal in UP for land belonging to scheduled tribes to be transferred to non-tribals.

The UP government has attempted to combat landgrabbing at various times, but in 1975 it regularized all land encroachments that had occurred up to June 30, 1966.[64] By the regularization, the government gave way to the landgrabbers' interpretation of the preferred tarai landscape and allowed their interventions in the physical landscape to stand. Afterwards, the UP government tried, with limited success, to enforce its interpretation of the landscape and evict landgrabbers from their holdings. In 1975, the government filed several court cases against landgrabbers, but Tharus and Buxas who had lost land rarely got it back since land lost before 1968 was not involved.[65] The UP government was still acting against landgrabbers in the tarai of Kheri district as recently as 1994.[66]

Political Violence and Banditry

One other important factor involved in the development of the tarai has been political violence and banditry. Prior to 1947, the tarai had the reputation of being the home of dacoits or bandit gangs. The most well-known of these dacoits before independence was one Sultana, who was described by Jim Corbett as "India's Robin Hood."[67] It should be noted, however, that banditry did not cease to exist in the tarai after 1947. Furthermore, the tarai has been the location for political violence since the 1970s.

As noted above, Arjan Singh described the activities of dacoits in Kheri district, but dacoits were also found in Pilibhit and Nainital districts. In 1953, the UP government was forced to take action against dacoits who operated in the colonization area of southern Nainital district. Just like the dacoits of the past, such as Sultana, the bandits of Nainital district used the forests and swamps as places to hide from the police. In April of 1953, the government of UP sent four Provincial Armed Constabulary platoons along with the district police into the colonization area to capture the bandits, but with limited results.[68]

There have been two forms of political violence in the UP tarai after 1947. The first is violence between landgrabbers and the state and between different groups of landgrabbers and their political allies. The second is related to separatist Khalistan violence in Punjab state in the 1980s. Indeed, journalist Saibal Dasgupta has written that "[w]hat was merely a spill-over from Punjab has now assumed the proportions of a full-blooded menace, with the Terai region becoming a 'mini Punjab' in terms of the scale of violence."[69]

It is crucial to note that political violence in the tarai has been committed by different actors including Sikh individuals and groups as well as local and state officials. For example, police officials have blamed militant groups like the Bhindranwale Tiger force of Khalistan for violence against both Sikh and Hindu civilians in Pilibhit, Nainital, and Kheri districts.[70] In 1990, the UP government claimed that Sikh "terrorists" killed six policemen and 47 other people.[71] Officials, however, themselves have been accused of illegal violence by journalists, Sikh political groups, residents of the tarai, and NGO organizations like Asia Watch and the People's Union for Democratic Rights. In 1992, journalist Saibal Dasgupta reported that "A police officer in Pilibhit ordered the execution of ten Sikhs who he dubbed terrorists. The state government's brazen defence of the action helped the extremists' propaganda about the BJP...being a party gunning for the Sikhs."[72] Asia Watch and the People's Union for Democratic Rights have also accused UP police of summary executions and other extra-legal actions.[73]

It should be noted that most, if not all, of the political violence between officials and Sikh individuals and groups occurred after the settlement of the tarai and the initiation of the "green revolution." The violence has arisen as part of the development of new social relationships in the tarai due to immigration and subsequent economic prosperity for some of the immigrants. Immigrants from UP, Bengal, and Punjab came to the tarai and there have been tensions between the different groups. As a group, Sikhs have benefited economically from the "green revolution," and this success has been disproportionate to their population in the UP tarai.[74] Saibal Dasgupta argues that "[t]he Sikhs here are a minority in demographic terms. But they carry economic clout as owners of the sprawling acres that have remained untouched by land ceiling laws..."[75]

By the end of the 1960s, the physical landscape of the tarai took the form of a patchwork as the terrain of the tarai was divided between different uses, including big and small farms and wildlife conservation. In addition, the new society that emerged in the tarai exhibited a similar heterogeneity as settlers from different parts of India formed separate colonies. Individual colonies were usually populated by relatively homogenous groups, like Bengali and Punjabi refugees, UP political sufferers, or ex-servicemen. Furthermore, the adivasi population of the tarai was never integrated, so that the Buxas and Tharus lived in separate villages from the Punjabis and Bengalis. The Buxas and Tharus tended to organize their villages according to their cultural norms as did the Bengalis. According to geographer Shailaja Pandey, "Tharus and even more Bokshas [sic] and Bengalis cling strongly to their traditional styles" in house construction and cropping patterns.[76]

A patchwork landscape was produced as different actors sought to realize their differing interpretations of the tarai. This process was intensified in the 1960s as the "green revolution" led to greater economic and social differences between the various groups that lived in the tarai. The tarai became a normalized part of India, but along different lines than those envisioned by the Tarai and Bhabar Development Committee and its chairman, G.B. Pant.

Notes

[1] Settlers who were allotted land through the official process had holdings from 10 to 50 acres. In contrast, in Kheri district Arjan Singh's farm, Jasbirnagar, occupied some 750 acres. In 1959, he leased an additional 173 acres, Tiger Haven, next to the North Kheri Forest Division.
[2] Singh claims that he was the first settler in the area after Indian independence. See Singh 1993a:54.
[3] Singh 1973a:29. Of course, Singh is describing a population suffering from the long term effects of malaria morbidity. As discussed earlier, it is all too easy to dismiss such populations as lazy instead of chronically ill.
[4] Shukla 1995:30.
[5] Singh 1973a:23.
[6] Singh 1973a:23.
[7] Singh 1993a:46.
[8] RRCD file number 1068/48, volume 2, p. 9.

[9] Singh 1993a:44.
[10] The fishing cat or *Felis viverrina* is a rare animal that is slightly larger than the domestic cat. As the name implies, the cat is adept at swimming and fishing. It is also known to attack goats, calves, and dogs. In the tarai it can be found in Dudhwa National Park. Breeden 1993:20-23.
[11] Singh 1993a:44 and Singh 1973a:31.
[12] Shukla 1995:32.
[13] Singh 1973a:29.
[14] Singh 1984:49. According to Singh 1973a:33 malaria throughout Nighasan had nearly been eradicated by 1959.
[15] Singh 1982:16.
[16] Singh 1982:16.
[17] Singh 1993a:54.
[18] Shukla 1995:35.
[19] Shukla 1995:152.
[20] Shukla 1995:152.
[21] PWCD file number 33/w/1951, volume 2, p. 1.
[22] PWCD file number 33/w/1951, volume 2, p. 4. According to CI, Series 22, Uttar Pradesh, Pilibhit District Census Handbook, Part xiii-a, p. 145, Neoria possessed some 125.05 hectares of land irrigated by tube wells.
[23] 1971 Census, Series 21, Uttar Pradesh, Pilibhit District Census Handbook, Part x-a, pp. 20-21. 286 acres were irrigated by canals and 291 acres were irrigated by electric tubewells.
[24] Singh 1982:12 and Singh 1973a:16.
[25] Singh 1973a:45.
[26] Singh 1973a:182 and Singh 1993a:55. He does not, however, name any names. Shukla 1995:193 charges that "terrorists, militants and naxalites" also engaged in poaching.
[27] Singh 1973a:19.
[28] Shukla 1995:34.
[29] Shukla 1995:34.
[30] Stracey 1957:575.
[31] Spillett 1967:623, 625.
[32] Gee 1992:159 and Spillett 1967:624.
[33] Ali 1985:187.
[34] Stracey 1957:576.
[35] A. Royds "India's Dwindling Wild Life" *The Pioneer*, June 20, 1954, Pioneer Magazine, p. 3, Mukherjee 1982:vi and Stracey 1960:18
[36] GOI, Indian Board for Wildlife 1970:7.
[37] GOI, Indian Board for Wildlife 1970:29.
[38] GOI, Indian Board for Wildlife 1970:32.
[39] Stracey 1960:24.
[40] RCD file number 41c/1948, p. 4.
[41] RAD file number 386/49, volume 1, p. 20.
[42] Pandey 1982:136.
[43] RCD file number 41c/1948, p. 111.
[44] Quoted in Randhawa 1980-86, 4:57.
[45] RAD file number 386/49, volume 1, pp. 20.
[46] RAD file number 386/49, volume 1, pp. 1-15.

[47] RAD file number 386/49, volume 1, pp. 14-16.
[48] RCD file number 127/1951, p. 1.
[49] RCD file number 127/1951, pp. 1-2.
[50] RCD file number 127/1951, p. 3.
[51] RCD file number 127/1951, p. 15.
[52] RCD file number 127/1951, pp. 17-88.
[53] Hasan 1976:11.
[54] Quoted in Hasan 1992:63.
[55] Hasan 1976:11 and Hasan 1979:194.
[56] Rawat 1993:98.
[57] Quoted in Hasan 1992:144.
[58] Rawat 1993:99 and PUDR 1989:3.
[59] Rawat 1993:97.
[60] Joshi:1969:6 and Misra 1993b:2059.
[61] PUDR 1989:10 and Misra 1993b:2059.
[62] Singh 1973a:53.
[63] Singh 1973a:53.
[64] Rawat 1993:101 and Hasan 1979:190-191.
[65] Hasan 1976:11, Hasan 1979:190-191 and Hasan 1992:145. See also Rawat 1993:101.
[66] *The Pioneer,* January, 24, 1994, p. 4.
[67] Corbett 1989: 90-131. See Smythies 1961:94 for a less romantic view of Sultana.
[68] See *The Pioneer*, "Dacoit Menace in Tarai Area" March 21, 1953, p. 2 and *The Pioneer*, "Dacoit Menace in Naini Tal Terai" April 9, 1953, p. 5.
[69] Saibal Dasgupta "Roots of Terrorism in Terai" *Times of India*, July 28, 1992, p. 8.
[70] *The Times of India*, "Punjab Militants Kill 29 in Pilibhit" August 4, 1992, p. 1; *The Pioneer*, "Huge Arms Cache Recovered in Lakhimpur-Kheri" December 27, 1993, p. 4; *Times of India*, "Terrorists Loot Bus Passengers" February 4, 1994, p. 1, and Sharad Gupta "Terai Terrorism on Wane—Indelible Scar on People's Psyche" *Times of India*, February 1, 1994, p. 1. See also Kaur 2000.
[71] The Economist 1991:24.
[72] Saibal Dasgupta "Roots of Terrorism in Terai" *Times of India*, July 28, 1992, p. 8. See also Bains 1992:1 and Misra 1993a:1421.
[73] See PUDR 1988, PUDR 1989, Asia Watch 1991, and Kaur 2000.
[74] Randhawa 1980-1986,4:51-60.
[75] Saibal Dasgupta "Roots of Terrorism in Terai" *Times of India*, July 28, 1992, p. 8. See also Misra 1993a:1422.
[76] Pandey 1982:139.

• CHAPTER SEVEN •

The Changing Tarai Landscape

After 1961, that is, after Pant's tenure as Chief Minister of Uttar Pradesh, there were several changes in the normalization of the tarai. The government altered its interpretation of the tarai, ceased its land reclamation program, and encouraged the development of more intensive methods of agricultural production, which resulted in the introduction of the "green revolution" into Uttar Pradesh. With the end of agricultural colonization, the state government slowly embraced a more active role in wildlife conservation and began to revise its assessment of what constituted "wasteland" and it began to see value in land that once was considered worthless, such as swamps or scrub land that served as wildlife habitat. Furthermore, when the UP government as well as the GOI began to see value in the wildlife itself, the preferred landscape of the tarai envisioned by both governments changed. Land devoted to use as animal habitat ceased to be interpreted as "waste" and was increasingly seen by government officials as sites for human recreation as well as wildlife conservation. A new preferred landscape began to emerge as government officials in Lucknow and New Delhi agreed that part of the tarai should be set aside for wildlife conservation. They reversed an earlier perception that wildlife was solely a menace to agriculture and began to argue that wildlife in itself was valuable. This shift in landscape preferences led to a change in the tarai's physical landscape, through a series of mediations, in six ways: the creation of the UP state Board of Wildlife in 1964, the expansion of Corbett National Park in 1966, the designation of the North Kheri Forest Division as a wildlife sanctuary in 1965, the passing of the Wildlife (Protection) Act of 1972, the initiation of Project Tiger in 1973, and the creation of Dudhwa National Park in 1977. These actions strengthened and extended wildlife conservation in the tarai by setting aside more land for conservation as well as increasing the legal protections of both the animals in the parks and the parks themselves. For example, the Wildlife (Protection) Act of 1972 banned all sport hunting in India and banned all timber activities inside national parks. This represents a major shift in the preferred tarai landscape that would influence changes in the physical landscape when, for example, legal forms of hunting were replaced by poaching.

By the 1960s, the idealized tarai landscape envisioned by Pant, Katju, and the Tarai and Bhabar Development Committee was not only partially realized, but it was beginning to be replaced by the preferred landscapes of others. Large and small farmers, conservationists, tribals, and others were competing for access to the resources necessary to implement their varying visions. The resulting patchwork landscape can be seen as a series of overlapping uses of the land where wildlife conservation sites bordered small subsistence-oriented farms, which in turn competed with large mechanized, commercialized farms.

In addition, beginning in the late 1960s, there was a sharp increase in the incidence of malaria morbidity after several years of decline throughout the tarai. This unanticipated change in the physical tarai landscape forced a revision of the government's late 1950s and early 1960s version of the tarai as "healthy" or at least relatively free from malaria. One consequence was the government's perception that the "green revolution" in the tarai was threatened by the upsurge in malaria and steps had to be taken to protect it.[1] The resurgence of malaria was a problem found throughout India and was in no way unique to the tarai or Uttar Pradesh. An increase in malaria morbidity occurred in several places throughout India, and even the world. Indeed, malaria has been a persistent problem in many parts of India (and the world) up to the present day.

Changing Agricultural Technology and the "Green Revolution"

With the end of the second Five Year Plan, the government of Uttar Pradesh (as well as the GOI) no longer sought to increase agricultural production through large-scale reclamation of cultivable wasteland. In the third Five Year Plan, instituted in 1961, and afterwards, the government stressed improved agricultural techniques and more intensive cropping as the means for increased food production. In the tarai, the UP government established the UP Agricultural University at Pantnagar (later renamed Govind Ballabh Pant University of Agriculture and Technology), a land-grant institution built on the grounds of the former UP State Farm in Nainital district, with the intention to provide "for the development of agriculture and for the benefit of the rural people of Uttar Pradesh."[2] The university, through its extension programs and scientific experiments, became the foundation of the "green revolution" in the tarai and western UP by the late 1960s.[3] The university served as not only a testing site for the new technology, but as a distribution point as well. The university distributed the new technology during educational training sessions with farmers and by serving as a High-Yielding Varieties [HYV] seed store.

The "green revolution" in India has been defined as the significant increase in agricultural productivity due to the use of wheat and rice hybrid seeds, chemical pesticides and fertilizers, and advanced agricultural techniques and machinery brought about by the GOI's High Yielding Varieties Programme.[4] While the tarai is an important, even central site for this process, the "green revolution" has proven to be a more complicated process. The very term "green revolution" implies that technological innovations such as hybrid seeds and chemical fertilizers were suddenly adopted in the mid-1960s and led to a swift and unprecedented increase in agricultural production. Prem S. Mann aptly represents this interpretation of the "green revolution" in India:

> Until the early 1960s the farmers in India were using obsolete and traditional modes of cultivation. There was very little use of modern techniques with inputs like tractors, threshers, combine harvesters, tubewells, pumpsets, fertilizers, pesticides, and insecticides.... By the mid-1960s, however, a spectacular change was being witnessed in the use of technology and inputs in Indian agriculture.[5]

In this rhetoric of the "green revolution," a rhetoric that was common among the government officials, scientists, and academics[6] involved in the early stages of the "green revolution," India had been mired in "traditional" agriculture until the overnight adoption of HYV seeds and other "green revolution" technologies.[7]

Furthermore, scientists at the Rockefeller Foundation, who played an important role in the introduction of "green revolution" technology into India, had argued in the 1950s that India had lost all facility for innovation and change. They further claimed that "[m]ore millions are enslaved by centuries of tradition and are not truly free to try new methods or to exploit their own ingenuity."[8] Foundation officers J.G. Harrar, Paul C. Mangesldorf, and Warren Weaver produced a report entitled "Notes on Indian Agriculture" that shaped Foundation thinking about India for the next two decades. In this report Harrar, Mangesldorf and Weaver argued that India was facing a food crisis which could lead to famine and political instability.[9] Furthermore, they argued that Indian agriculture was backward and mired in centuries-long tradition. India's farmers had not been affected in any way by the country's agricultural research programs.[10] They concluded that

> The most serious problem faced by agriculture in India is not a technical one, but a cultural one. The greatest handicaps which agricultural development must overcome are those imposed by the caste system, by ignorance, by religious prejudices, by

multiplicity of languages, by the many stifling customs, and by habits of thought which prize tradition over improvement.[11]

Such calumnies failed to recognize that Indian agriculture was not static prior to the 1960s.

In fact, there is little that was revolutionary in the "green revolution" technology adopted in Uttar Pradesh. B.H. Farmer has warned that "it is wrong to overemphasize the revolutionary aspects of "the new technology" on South Asian agricultural production as though agriculture in the subcontinent were sunk in a static stupor before the arrival of that technology."[12] In South Asia generally and in the tarai specifically, agricultural research, experimentation, and technological innovation have a long history. Furthermore, the adaptation and adoption of new technologies must be seen in the context of the simultaneous commercialization of agriculture. As new technologies were introduced into UP, agriculture ceased to be primarily subsistence-based. New crops, including cash crops, were introduced into the state, and these often required the importation of new inputs, like fertilizers and hybrid seeds. Significantly, the new technology was expensive to acquire and use. A subsistence farmer who had a two acre holding could not afford to purchase or use tractors, petrochemical fertilizers, or tubewells. Only farmers with access to relatively large amounts of capital could participate in the new commercialized agricultural economy. The effect of this split was a substantial change in the social relations of agricultural production in the tarai even before the mid-1960s.[13]

While the "green revolution" was the adoption of specific technologies by farmers in the 1960s, it was only one more step in a long process of agricultural innovation in the area. Agricultural experimentation and the use of new agricultural technology began in the nineteenth century on the Tarai and Bhabar Government Estates [TBGE]. The TBGE administrators in the nineteenth century introduced technological change when they constructed a variety of drainage and irrigation works in the Estates. The government of UP, by introducing canal irrigation works, did the same throughout much of western UP in the same period. In the twentieth century the TBGE administration sponsored experiments with steam plows and improved agricultural techniques. Furthermore, the state farm at Khatima not only conducted experiments, but also maintained a seed store that distributed improved (hybrid) seeds to Estates tenants. The Estates administration introduced cash crops like cotton and sugarcane as well, and it constructed a sugar-processing plant in Khatima. While this should not be taken to mean that technological innovation was evenly distributed throughout the TBGE or even particularly effective, it suggests that agriculture in the region was not static and unchanging prior to 1965.

Agricultural research and experimentation occurred not just in the TBGE, but throughout India before independence. While most of this research did not affect the lives of most farmers, a research infrastructure was being created. This means that the research facilities associated with the "green revolution" did not magically appear overnight in the 1960s. In the nineteenth century, the GOI had sought to increase agricultural production (and the revenue derived from it) through the expansion of canals and irrigation works.[14] The expansion of irrigation not only increased the amount of land in production, it also led to an increase in the cultivation of cash crops, like sugar cane, indigo, and opium.[15] The GOI created its first Agriculture Department in 1870, and the government of the North-Western Provinces (a forerunner of the United Provinces of Agra and Oudh) followed suit in 1874. These departments promoted experimentation with imported food grains. They also promoted trials of US-made windmills and the steam plow in the 1880s.[16] The GOI began to more actively support agricultural research and technological innovation with the creation of the Imperial Veterinary Research Institute at Mukteswar in 1893 and the Imperial Agricultural Research Institute at Pusa in 1902.[17] These institutions became major sites for agricultural research, including the development of new seed hybrids.

In 1905, the GOI further intensified its support of agricultural research. According to Carl E. Pray, "[a]gricultural research started on a systematic basis in British India with the reorganization and re-establishment of the Department of Agriculture in 1905. It was quite successful in developing improved varieties of crops."[18] Afterwards, the GOI was disappointed with the level of agricultural productivity in the country, so in 1926 it constituted a Royal Commission on Agriculture. In its 1928 report, the Commission made several recommendations, one of which led to the creation of the Imperial Council of Agricultural Research. As part of its contribution to the proceedings of the Commission, the UP government pointed out that its department of agriculture ran a number of farms that bred improved quality livestock and grew and distributed improved seed. The UP government argued that one of the most serious problems in Indian agriculture was not a lack of agricultural research, but the effects of malaria and other diseases on the rural population.[19] In addition, as part of the Grow More Food Campaign in the 1940s, the GOI and the government of UP jointly funded an extension program in the Kumaun and Rohilkund areas of UP, which included Nainital and Pilibhit districts; this program distributed improved rice seed to farmers.

The program also provided for the demonstration to farmers of new agricultural machinery.[20]

The GOI continued to support agricultural of research after independence, especially through the facilities of the Indian Council on Agricultural Research and the Indian Agricultural Research Institute. The GOI's involvement with agriculture changed with the adoption of the first Five-Year Plan in 1951 and a gradual de-emphasis on the Grow More Food program in the 1950s. The first Five-Year Plan was concerned mostly with repairing the damage caused to Indian agriculture by World War II and partition.[21] Furthermore, several of the state governments became active in agricultural research in the late 1950s and early 1960s. As part of the state's first five-year plan, Uttar Pradesh in 1951 created an extension service as part of the Department of Agriculture, with the stated purpose of educating farmers.[22] Uttar Pradesh was the first state to create an agricultural university, passing the necessary legislation in 1958. The UP Agricultural University, the first land-grant university in India, supported agricultural research and provided extension services to farmers in the tarai and western UP.

Historian S.K. Mukherjee argues that there was an increase in the production of food crops in India even before the beginning of the "green revolution." Mukherjee suggests that this increase was due to improved seeds and agricultural techniques plus the increased use of irrigation and fertilizer.[23] According to the Indian Council on Agricultural Research, some 28,932 field experiments were conducted in India from 1948 to 1959 and roughly one-third of them were conducted in Uttar Pradesh.[24] Agricultural research in the tarai at this time was conducted at four separate research stations located on the Tarai State Farm as well as at the Soil Research Laboratory and the Regional Research Station, both in Rudrapur town in Kichha tahsil, Nainital district.[25] Brian Lockwood argues that the term "green revolution" is "inappropriate" because "[t]he factors which have localised the 'Green Revolution' are part of a broader change in farm technology which has been proceeding for decades in the better-endowed farming areas" of the Punjab, Haryana, and western UP.[26] Clearly, Indian agriculture was not static prior the "revolutionary" 1960s, even though the level of technology available to and utilized by farmers varied greatly by class and location in the country.

The "green revolution" in India is best defined as the adoption of certain, specific mid-1960s technologies, not as some unprecedented break with "traditional" agriculture. The most important facets of "green revolution" technology are HYV hybrid seeds and the farm machinery, irrigation facilities, fertilizer, herbicides, and pesticides necessary to utilize them.[27] The HYV hybrid seeds and fertilizer, while important, are part of a larger "green revolution" technology "package."²⁸ Hybrid seeds were first developed in the United States in the 1910s and 1920s, and HYV seeds are particularly productive strains of hybrids.²⁹ HYV seeds were first developed in the 1950s and 1960s through research sponsored, in part, by the Rockefeller Foundation and the Ford Foundation in Mexico, India, and the Philippines.³⁰ In 1956, the Rockefeller Foundation and the GOI signed an agreement that created the India Agricultural Program which funded research programs throughout the country. Thus the Rockefeller Foundation played a leading role in the transfer of information about agricultural research methodology to India in the 1950s and 1960s.³¹ Significantly, the Rockefeller Foundation became involved in Indian agricultural research at the request of the GOI. The GOI Secretary of Agriculture Vishu Sahay contacted the Rockefeller Foundation in 1953 for its assistance with maize research. Maize was the crop of choice because the GOI felt that its wheat and rice research programs were reasonably successful and did not need foreign assistance. ³² This "mutuality of interest" made it possible for the GOI and the Foundation to work successfully, so that, in short, the Foundation's efforts in India lacked somewhat the know-it-all neo-imperialist approach found in Foundation projects in other countries.³³ The Ford Foundation, the US-based Agricultural Development Council, and the US Technical Cooperation Administration, predecessor to the US Agency for International Development, also funded agricultural research in India during the 1950s.³⁴ The Indian Agricultural Research Institute, relocated to New Delhi after 1947, was the leading Indian institution for agricultural research at this time and was the site for much of the agricultural research conducted in India until the development of the country's land-grant agricultural universities in the 1960s. Researchers in the late 1950s and early 1960s focused on developing new hybrids of rice, wheat, and maize, and these HYV seeds became available to farmers beginning in the early 1960s.³⁵ However, while Indian researchers adapted the HYV hybrids to suit local conditions, the first wheat and rice HYV hybrid seeds themselves were developed outside of India.³⁶ In 1963 the GOI created the National Seeds Corporation which had the responsibility of producing and distributing improved hybrid seeds.³⁷

In response to a Ford Foundation report entitled "India's Food Crisis and Steps to Meet It," the GOI created a program to further the development and distribution of the new hybrid seeds. In 1959, the Intensive Agricultural Districts Program [IADP] was established jointly with the Ford Foundation.³⁸ The IADP selected areas which had an adequate supply of rainfall or irrigation and which promoted the increased use of inputs like fertilizers. The IADP proved to be less than entirely successful and was canceled in the mid

1950s.[39] The GOI and the Rockefeller Foundation determined that the local improved seed varieties were not sufficiently responsive to the higher use of fertilizer to justify the expense.[40]

In 1965 the GOI announced a "New Strategy" to increase agricultural production, and it included the High Yielding Varieties Program [HYVP] which was a joint effort between the GOI and the Rockefeller Foundation and included the development and testing of new hybrid varieties of wheat and rice.[41] The New Strategy was a larger version of the IADP. Like the IADP, it sought to promote the increased use of fertilizers, but it also offered farmers with new hybrid varieties which were more responsive to the higher amount of fertilizer used. In addition to promoting new hybrid varieties and greater use of fertilizer, the New Strategy also included the investment in fertilizer factories in India, the extension of adequate credit to enable farmers to use the new varieties, the control of grain purchase prices to give farmers the financial incentive to try the new varieties, and the reorganization of agricultural research in India.[42] The National Seeds Corporation was given the primary responsibility of distributing the new hybrid HYV seeds to farmers, but state departments of agriculture also were responsible for the multiplication and distribution of the new hybrids.[43]

Analysis of the New Strategy by policy makers, scholars, and journalists began immediately. In 1968, agronomist V.S. Vyas concluded the New Strategy had proven to be more productive than "traditional agriculture." Vyas, however, noted that the program faced several difficulties, such as a lack of coordination between the various government agencies involved and the failure of state extension programs and agricultural cooperatives to adequately support the program. Vyas concluded that the program could become a complete success if these bureaucratic questions were addressed.[44] In 1969, Ralph W. Cummings and S.K. Ray argued that the New Strategy had had a profound effect on grain production. They acknowledged that the New Strategy had problems, such as they unreliable availability of inputs and the instability of grain prices, but they suggested that these problems could be resolved.[45] In 1969, Francine R. Frankel argued that the New Strategy was based on false assumptions. Most importantly, Frankel argued that the new HYV seeds were not any better than the older hybrids already in use and thus the New Strategy would produce, at best, mixed results.[46] In a 1976 study by the GOI's Programme Evaluation Organization and the Australian National University researchers concluded that, despite problems, the New Strategy and HYVP had "substantially increased production of foodgrains in India."[47] Other commentators argued that the New Strategy was not as successful as its proponents claimed. They suggested that it was hampered by administrative problems, shortage of agricultural inputs, especially fertilizer, lack of adequate irrigation facilities, social and economic instability, and environmental problems, such as the chemical exhaustion of the soil and the salinization of the soil.[48]

Agricultural universities throughout India were central to the development, distribution, and promotion of new hybrids, fertilizers, pesticides, and agricultural techniques in the 1960s and 1970s. As noted above, the UP Agricultural University at Pantnagar in Nainital district was the first land-grant or agricultural university in India. The idea of creating agricultural land-grant universities in India originated with the GOI's University Education Commission as well as the GOI's Damle Committee on Agricultural Research and Education. Harpal Singh Sandhu, UP Assistant Director of Colonization, invited the Damle Committee to visit the Tarai State Farm and consider it as the site for an agricultural university. The Committee liked the site and recommended that an agricultural university be built there. Sandhu and A.N. Jha, UP Food Production Commissioner, were sent by the Committee to the US to visit several land-grant universities in 1950, and they recommended to G.B. Pant that an American-style agricultural university should indeed by built on the grounds of the Tarai State Farm. Pant approved, and the state of UP made a proposal to the GOI in 1956 to create an agricultural university.[49] The GOI approved the proposal, and the UP legislature then passed the Uttar Pradesh Agricultural University Act in 1958. With GOI approval, the UP government received funding to begin construction through the Indo-US Technical Cooperation Programme.[50] The university was sited on the grounds of the former UP State Farm in Nainital district and Prime Minister Jawaharlal Nehru inaugurated it in ceremonies held on November 17, 1960.[51] The University of Illinois at Urbana-Champaign was involved in the development of the university from 1960 onward as Illinois university architects assisted in the original construction of the Pantnagar campus as well as later additions.[52] The farmland of the UP State Farm was given to the university as both as source of revenue and as a site for agricultural research and experimentation.[53] G.B. Pant died in 1961 and the UP Agricultural University at Pantnagar was renamed in his honor the Govind Ballabh Pant University of Agriculture and Technology in 1970.[54]

Pantnagar University soon became a significant force in the development and transfer of HYV seeds and related technology to the tarai.[55] The University promoted the adoption of HYV technology by farmers in two ways. Firstly, the University sought to educate farmers about HYV technology through degree courses on campus and through an extension system that sent extension agents to visit farmers and demonstrate the new technology.[56] As part of the educational effort, the University created four regular publications: the annual report for the University's experiment station, the *Indian Farmer's Digest*, the *Indian Agricultural Index*, and the *Pantnagar Journal of Research*. It is not clear how much influence these publications had on tarai farmers, however, because only the *Indian Farmer's Digest* had a Hindi language edition. Secondly, University scientists conducted research in all aspects of agriculture, including agronomy, entomology, horticulture, veterinary medicine, irrigation technology and methodology, and grain storage technology. More specifically, the University developed and tested new seed hybrids, fertilizers, and pesticides specifically for use in the tarai. The University has produced new maize, wheat, and rice varieties, but, while maize, wheat, and rice have been the core of the "green revolution,"

The US greatly influenced the development of the University through its funding of several University programs. Many of the University's research activities were funded by the Rockefeller Foundation, the Ford Foundation, the Indo-US Technical Cooperation Programme, and the US Agency of International Development (program Public Law 480). The University also received funding from student tuition, income from the operations of university farm, and various external agencies, including the Indian Council of Agricultural Research, the GOI Planning Commission, the GOI University Grants Commission, the UP State Council of Scientific and Industrial Research, the Bhabha Atomic Research Centre (Bombay), and the World Bank.[57] The bulk of the university's operating budget, however, has come from the profits of the University farm.[58]

Pantnagar University served as a valuable resource for those farmers who wished to adopt HYV technology, but the availability of the necessary information and inputs was only one of several factors that affected the spread of the new methods in the tarai. There were several other important factors, including the availability of irrigation facilities and electricity, the availability and price of inputs such as seed, fertilizer, and pesticides, the availability of credit, and grain purchase prices. In short, farmers had to be convinced that the new technologies would be profitable before they would use it. Uncertainty about the profitability of HYV technology led many farmers, especially those with small holdings, to cautiously experiment with it. They would often devote part of their land to the new seeds while planting the remaining land with *deshi* varieties.[59] Furthermore, many farmers continued to cultivate *deshi* varieties because of periodic shortages of inputs like HYV seeds and fertilizer.[60]

The "green revolution" intensified the commercialization of agriculture in India in general and in the tarai in particular. The commercialization of farming represents a substantial shift in the preferred tarai landscape away from the small holdings and cooperatives envisioned by Pant and Katju toward the large, mechanized farms. One aspect of this is commercial and involves an increase in the cultivation of crops like cotton and sugarcane for cash.[61] Another is sociological and involves the question of participates in the "green revolution" in the first place. To participate a farmer must be able to purchase HYV seeds and the associated inputs, such as petrochemical fertilizers, herbicides, and pesticides. A farmer must possess the necessary capital or have access to adequate credit and many small farmers were unable to make many investments in HYV due to high interest rates as well as high input costs.[62] HYV crops require large quantities of water, so the farmer must have reliable access to irrigation. Furthermore, farm machinery like tractors, seed drills, and threshers are necessary to effectively use HYV seeds and related inputs. All of the equipment and inputs require a great outlay of capital, so much so that farmers who wish to use this technology must possess capital or have access to credit. Not surprisingly, participation in the "green revolution" by farmers in the tarai has been mixed; not all farmers have had equal access to credit, and only a portion have had holdings large enough to support "green revolution" technology.[63] As a result, there has been a marked shift in agrarian social relations as well as in the physical tarai landscape, and while some farmers have been able to fully participate in the "green revolution," others have not.

In the tarai, "green revolution" technology has been most widespread in Nainital district, perhaps because of its location near the UP Agricultural University at Pantnagar, which has provided farmers with easy access to HYV seeds and inputs.[64] Agricultural economists from Pantnagar have identified two categories of farmers: "progressive" farmers who have access to the resources necessary to utilize HYV technology and have devoted a significant portion of their acreage to HYV seed, and "less progressive" farmers.[65] Furthermore, farmers with larger holding have been found more likely to adopt HYV technology. This means that members of the group of big farmers, who owned the mechanized farms which emerged in the mid 1950s, were more likely to become "progressive" farmers.[66] In addition, those farmers who possessed the necessary capital and irrigation facilities to invest in HYV technology earned relatively larger incomes and were then able to invest further in HYV technology. An additional resource to consider was electricity which could be used to power the machinery used in the irrigation and cultivation of HYV crops. "Progressive" farmers tended to have greater access to electricity than "less progressive" farmers.[67] Alternatively, those farmers who lacked the resources to invest in HYV technology tended to earn lower incomes from their *deshi* varieties of crops. Thus they found it relatively more difficult to make the transition from *deshi* varieties to HYV seed.[68] As a consequence, farmers who possessed more than four hectares of land were significantly more likely to own farm machinery like tractors and also more likely to have reliable access to irrigation and electricity.[69] This process was intensified when the price of inputs, like seed, insecticide, fertilizer and pesticide, underwent a significant increase from 1967-68 to 1970-71. With the price of insecticides increasing by 50 percent or more, some small farmers found it difficult to invest in HYV technology.[70] Many large farmers, however, were able to generate a profit from the cultivation of wheat, rice, and maize despite the increased cost of inputs.[71] "Progressive" farmers were also more likely to cultivate newly introduced cash crops like soybeans and sugar beets.[72] That is not to say that small farmers never made any investment in HYV technology. Small farmers did borrow capital to make investments in HYV technology, but those farmers with large holding were more likely than farmers with small holdings to make such investments. Another factor to consider is that small farmers had difficulty obtaining loans because of the high interest rates charged to small farmers without collateral.[73]

By the mid 1970s, the tarai ceased to an exotic land that stood apart from neighboring district. The "green revolution" had altered the physical landscape away from the preferred landscape envisioned by the TBDC, and the tarai now shared many characteristics with western UP, Haryana, and Punjab where the "green revolution" had also been adopted widely. Indeed, tarai farmers faced the same difficulties regarding the sustainability of HYV technology as farmers elsewhere in India. These problems included the continued presence of crop diseases, rats, inefficient use of inputs, especially fertilizers, waterlogging and salinization of the soil due to inefficient irrigation, and the exhaustion of the soil from continuous cultivation.[74] It should be noted, however, that farms in the tarai were, on average, the largest in UP.[75]

At this time, agriculture in the tarai had been developed to the same levels as in neighboring districts. On average, farms in Pilibhit and Kheri had roughly the same level of mechanization and electrification as districts in central and eastern UP, like Basti, Etawah, and Shahjahanpur. The level of farm mechanization and electrification in Nainital was comparable to that of the most developed districts in UP, such as Muzzaffarnagar and Meerut while the average level of farm mechanization and electrification in Muzzaffarnagar, Meerut, and Nainital was lower than that in the Punjab and Haryana.[76]

Agriculture in the Tribal Areas of the Tarai

The implementation of the New Strategy was not conducted at the same rate throughout the tarai. Buxa, Tharu, and Bhotiya farmers did not adopt the new hybrids at the same rate as their immigrant neighbors.[77] The Bhotiyas, who were allotted land in the tarai from 1952-1959, lacked irrigation on their farms and were thus unable to utilize the new hybrid varieties, which require more water than older varieties.[78] Bhotiya farmers were unwilling to invest in the new technology because of the lack of irrigation on their farms.[79] Tharu and Buxa farmers as well were less willing than their immigrant neighbors to invest in the new hybrid varieties, petrochemical inputs and farm machinery. They were, however, interested in investing in irrigation facilities, which were a prerequisite for adopting the new varieties.[80] Furthermore, the Tharus and Buxas engaged mostly in subsistence agriculture and showed little interest in raising cash crops like sugarcane.[81] The adivasi economy did change after 1948 in that the Buxas and Tharus lost the customary access to forest produce that they had prior to the tarai colonization scheme. They then had to purchase wood from the Forest Department at the full commercial rate.[82]

In the early 1970s, the G.B. Pant University of Agriculture and Technology sent extension workers to demonstrate HYV technology to adivasi farmers in Nainital district.[83] As soon as 1974, however, economists from Pantnagar declared that the Tharus and Buxas "are completely unresponsive to modern innovations in farming resulting in their economic backwardness."[84] This perceived lack of interest by the tribals in adopting HYV technology reflects the age-old stereotype that the Buxas and Tharus are "lazy" and "fickle." Instead, the Buxas and Tharus lacked the capital to invest in HYV technology and were reluctant to borrow the money. This reluctance to borrow may be a result of one of the tactics used by landgrabbers to part adivasis from their land: indebtedness.[85] The adoption of differing levels of New Strategy technology illustrates the differing amounts of resources available to the immigrant farmers and tribal farmers because the adivasis lacked the resources available to the immigrants.

The UP government did attempt to improve the economic and social condition of the adivasis by establishing a number of economic development programs for scheduled tribes in the 1970s. The first such program was the establishment of a "Special Area Project" in the Nainital tarai town Khatima in 1970. This program involved the appointment of eight Village Level Workers, an Assistant Development Officer, and one Trained Midwife.[86] In 1975, the UP government established the Tarai Anusuchit Janjati Vikas, Ltd. [Tarai Scheduled Tribe, Ltd.], based in Lucknow, as part of the government's Directorate of Tribal Development. This corporation ran several programs in the eastern tarai, including Kheri district, which involved education, economic development, expansion of irrigation facilities, and marketing the surplus produce of tribal farmers. The corporation also was charged with conducting anthropological research amongst the tribals.[87]

The adivasis of Nainital were included under the jurisdiction of the Kumaun Anusuchit Janjati Vikas Nigam [Kumaun Scheduled Tribe Development Corporation], based in Nainital town. This group, a subsidiary of the Divisional Development Corporation, ran programs dealing with education and training, agricultural subsidies, and provision of raw supplies for tribal cottage industries.[88] In 1976, the UP government established an Integrated Tribal Development Project [ITDP] in the tarai area of Kheri district. This ITDP was different from other ITDPs in the state due to the absence of other development programs in the area and included economic development workers, medical personnel, veterinary workers, and teachers. The ITDP was the basic economic development unit of the government in the area, but was also charged with assisting the tribals with the loss of their land through landgrabbing. In 1980, the ITDP in Kheri was transferred to the jurisdiction of the Tarai Anusuchit Janjati Vikas, Ltd.[89]

Wildlife Conservation

The GOI formally took an active interest in wildlife conservation in 1972 with the passage of the Wildlife (Protection) Act 1972; but before this Act, wildlife conservation had been primarily a state subject. The UP government was supportive of wildlife conservation after 1947, but this interest was relatively low key. The state became more actively involved in wildlife conservation with the creation of the UP state Board of Wildlife in 1964, the establishment of the North Kheri Forest Division as a wildlife sanctuary in 1965, and the expansion of Corbett National Park in 1966.

New interest and activity in wildlife conservation was generated in India when the International Union for the Conservation of Nature [IUCN] held an international congress in New Delhi in 1969 at the invitation of the GOI. Three results of the renewed governmental interest in wildlife generated by the IUCN meeting were the banning of the export of tiger skins in 1970, the passage of the Wildlife (Protection) Act 1972 and the creation of Project Tiger in 1972.[90] Another step taken by the GOI and the UP government for the protection of wildlife was the creation of Dudhwa National Park and Kishanpur Wildlife Sanctuary out of the North Kheri Forest Division in Kheri district in 1977.

After 1947, wildlife conservation was a low priority in India. Most conservation activities consisted of reactive defense of existing conservation projects, like the preservation of Hailey National Park in the late 1940s. This is not to say that there was no constituency in support of wildlife conservation. Throughout the 1950s and early 1960s there was a great deal of discussion and debate regarding wildlife conservation, especially within the Forest Department. One feature of this was the creation by the GOI of the Indian Board for Wildlife in 1952. While the creation of the Board signaled a renewed interest in wildlife by the GOI, it had little practical effect because the Board had little legal authority. The Board was given the task of "devising ways and means of Conservation and control of Wild Life through co-ordinated legislation and practical measures and to sponsor the setting up of National Parks and Sanctuaries."[91] The Board, however, lacked the power to enforce its recommendations and the states were free to ignore them.[92]

The UP government followed the lead of the GOI by creating its own Board of Wildlife in 1964. In 1966, the area of Corbett National Park was increased to 523 square kilometers, but this increase was temporary. Since 1964, the UP government had been preparing to dam the Ramnagar river just outside of the park's boundary and, when this dam was completed in 1979, 18 square miles or about 10 percent of the park would be inundated.[93] According to Guy Mountfort of the Worldwide Fund for Nature [WWF], Corbett National Park was "partly spoiled" by the dam.[94]

In November 1969, the International Union for the Conservation of Nature held its tenth annual conference in New Delhi at the invitation of the GOI.[95] One result of the meeting was that the Indian (or Bengal) tiger was listed in the IUCN's Red Book as an endangered species.[96] Karan Singh, GOI Minister of Tourism and Civil Aviation and Chairman of the Indian Board for Wildlife, attended the conference and immediately afterwards persuaded Prime Minister Indira Gandhi to convene an Expert Committee to consider the protection of wildlife in India which reported to the Prime Minister personally.[97] In addition, the Board for Wildlife commissioned a census of the tiger population which determined that there roughly 1,800 tigers in India.[98] This number was lower than expected and alarmed conservationists in India and those affiliated with the WWF and IUCN.[99]

The Indian Board for Wildlife's Expert Committee released its report in 1970. The committee examined the various causes for the decline in all wildlife populations, not just the tiger.[100] In addition, the committee made a number of recommendations including the immediate permanent banning of the hunting of tigers and other endangered species and a three-year ban on the shooting of all animals and birds.[101] The committee further recommended a wide ranging reorganization of wildlife management throughout the country.[102] The committee determined that "[t]he wildlife of India is approaching extinction. This fact emerges inescapably from all reports."[103] The committee urged that action be taken immediately.

The GOI responded with the passage of the Wildlife (Protection) Act, 1972 and the creation of Project Tiger in 1972. The Act banned all sport hunting in India, but the legal shooting of tigers did not entirely stop. Tigers continued to be shot as part of crop protection activities, especially as the habitat of the tiger shrank. More importantly, the Act represented a shift in conservation strategy. Prior to the Act, wildlife conservation had taken primarily the form, with some exceptions, of the protection of individual species. The Act now established the principle of protecting an entire ecosystem or biotope and thus protect all the animals and plants in the area.[104] Project Tiger was based on this principle with the idea that the preservation of the tiger could only be successful if the tiger's habitat was preserved. According to Project Tiger official H.S. Panwar, the "[t]iger being at the apex of a complex food chain in most of our forest ecosystems, its status symbolises the state of health and the functioning efficacy of these ecosystems."[105] Project Tiger was a separate scheme overseen by the Indian Board for Wildlife, but the responsibility for the execution of the project was divided between GOI and state officials.[106] The government took several steps to protect the tiger's habitat, including greater law enforcement against poaching and the removal of 33 villages from the interior of Project Tiger reserves.[107] Initially, nine areas were chosen as Project Tiger preserves, including Corbett National Park, and other areas, like Dudhwa National Park were added later. The GOI has received assistance for Project Tiger from the United States Fish and Wildlife Service and the WWF, which collected donations worldwide in support of tiger conservation.[108]

Corbett National Park was selected as the first Project Tiger site and was dedicated as such on April 1, 1973. Upon becoming a Project Tiger reserve, the park was divided into two parts: an inner core or *sanctum sanctorum* and an outer periphery or buffer zone. Within the *sanctum sanctorum*, human activity was minimized and all timber operations and cattle grazing were halted.[109] In the buffer zone, some forestry activity continued, mostly the exploitation of minor forest produce.[110] No villages were removed from within the boundaries of Corbett, but two villages located near the park's border were relocated.[111] Corbett National Park has been one of the most notable Project Tiger sites and has one of the highest tiger densities in the country.[112]

The North Kheri Forest Division was declared a wildlife sanctuary in 1965 and the northern portion was reclassified as Dudhwa National Park in 1977 by Indira Gandhi during the Emergency.[113] Gandhi had first suggested that the forest be made a national park in 1973, but the UP government did not respond.[114] These changes in the status of the North Kheri Forest Division were the culmination of a long campaign led by conservationist Arjan Singh whose farm, Tiger Haven, bordered the forest north of the town of Pallia, tahsil Nighasan. Singh himself has observed that "[t]hanks to the avid and virtually single-handed lobbying by the present writer and the firm conviction of a conservation-minded prime minister, the late Mrs. Indira Gandhi," Dudhwa National Park was created.[115] Gandhi was a keen conservationist and used the conditions of the Emergency to create Dudhwa outside of the normal political and bureaucratic channels.[116] She created the park despite the opposition of the UP Chief Minister as well as the UP Forest Department. As a consequence the government of UP has not considered itself obligated to protect the park. Furthermore, the hostility of UP government officials to Dudhwa has meant that they have not been willing to establish a buffer zone around the park. This would require the removal of many farms and villages and the GOUP has refused to do it. The irregular creation of the park has also led to enduring anti-park sentiment among the people living next to the park.[117] Dudhwa was not included as a Project Tiger reserve until 1988 due to the opposition of the first Project Tiger director, Kailash Sankhala.[118] Sankhala has noted that "[Arjan] Singh kept up constant pressure on us to include it in Project Tiger....Singh succeeded after I left" in getting Dudhwa included in Project Tiger.[119] Despite its status as a national park, Dudhwa has been the site of a number of forestry operations, which were conducted at the behest of the Forest Department.[120]

Dudhwa National Park has been the site for the reintroduction of tigers, leopards, and rhinoceroses. Conservationist Arjan Singh conducted experiments, with the support of the Indira Gandhi, in the reintroduction of three leopards and one tiger in the 1970s.[121] Singh's reintroduction efforts were controversial, but the most debated topic was the reintroduction of Tara, a zoo-bred tigress born in the United Kingdom. Critics charged that a tigress born in a zoo would never take to the wild, but they also claimed that Tara was not a pure-bred Indian tiger, but a cross between different tiger subspecies and hence a genetic "cocktail" which would corrupt the purity of the Indian tiger.[122] Kailash Sankhala dismissed Tara as an ecological as well as genetic mistake.[123] Singh has steadfastly argued that the reintroduction of Tara was an unqualified success because Tara returned to the wild and lived for 10 years. Project Tiger experts, however, have concluded that Tara died sometime shortly after her reintroduction in 1978.[124]

In 1981, the Indian Board for Wildlife, with the support of Indira Gandhi, began to discuss the reintroduction of rhinoceroses into Dudhwa National Park. The board felt that the park's "swampy grassland habitat" was ideal for the rhinoceros and six were in brought to the park from Assam in 1984.[125] In 1985, four rhinoceroses obtained from the government of Nepal in a trade for 16 elephants were brought to the park.[126]

The tarai of Kheri district, including the area of Dudhwa National Park, has been the site for great controversy regarding the interpretation of the preferred and physical tarai landscapes. The areas bordering the park have seen a large number of incidents of both man-eating and cattle-lifting by tigers since the beginning of agricultural colonization. The first three incidents of man-eating by tigers near Dudhwa occurred in the northern part of Nighasan district from 1959 to 1961. Tigers and humans were brought into close proximity due to the reclamation of land for agriculture in an area bordering the North Kheri Forest Division. These three tigers were shot by Arjan Singh.[127] There are several factors involved in man-eating and cattle-lifting: the failure to organize Dudhwa into a core-periphery structure found in other Project Tiger reserves, the existence of farms adjacent to the park's boundaries, the decline in the population of prey species through poaching and loss of habitat, and the politics of agricultural colonization. Singh insists that the familiarity of tigers with humans resulting from frequent human-tiger interaction was one of the factors involved in the cases of man-eating after 1947 and he concludes that "TIGER AND MAN CANNOT SHARE SPACE [sic]."[128] Dudhwa, however, was not the only national park where the core-periphery structure was not firmly established. By 1991, eight Project Tiger sites contained villages in their core areas, and nine of them had villages in their periphery or buffer zones with most of the villages populated by adivasis or other forest-dependent populations.[129]

The incidents of man-eating by tigers did not resume until 1978 when over 200 hundred people were killed by several tigers over a ten year period. The return of man-eating tigers coincided with the reintroduction of the zoo-bred tigress, Tara, by Arjan Singh into Dudhwa National Park. Several critics of Singh have argued that Tara was unable to behave as a wild tiger due to her life-long familiarity with human beings and that she soon turned to man-eating after being released into the forest. These critics that Tara was responsible for the rash of tiger attacks on humans and also claim that she was shot in 1980.[130] An investigation by Project Tiger officials, however, cleared Tara of man-eating charges, but concluded that Tara died shortly after her reintroduction into the forest in 1978.[131] This investigation also identified several tigers which had engaged in man-eating.[132] According to Rahul Shukla, there was controversy over the exact number of man-eating tigers in Nighasan.[133]

The high incidence of man-eating by tigers in Kheri district, rivaled only by the attacks on humans in the Sundarbans in Bengal, was due to the high rate of human-tiger interaction in those areas bordering Dudhwa National Park.[134] There has been a great deal of human-tiger interaction because of the patchwork of farms and forest in the tarai of the district. Dudhwa lacks a buffer zone and farmers have extended their sugarcane fields to forest's edge.[135] The sugarcane fields then provide cover for wild animals:

> the sugarcane fields attract wild ungulates from the Park, followed by tigers. Since the typical cultivation regime of sugarcane involves long periods of tranquility and relatively little human activity in the fields, tigers tend to become resident and to breed there, as well as preying on any wild ungulates or cattle in the area. In such a situation there is a high possibility of confrontations between tigers and people entering the fields.[136]

To further complicate the situation, unlike at Corbett National Park, villages located next to Dudhwa were not removed when the forest was declared a national park and a Project Tiger reserve. From 1977 to the early 1990s, the number of villages located near the park's boundaries grew from 21 to 80. Not only do the fields of the village farmers extend to the park's edge, but the villagers rely on the park to meet their fodder and firewood needs.[137] Rahul Shukla argues that "after India's independence and specially in the last decade, many man-eaters have been products of biotic disturbances, for many of its areas have been thoughtlessly ravaged by the ploughers and harvesters."[138] This situation represents a clash over the interpretation of the tarai landscape. Indira Gandhi and the Project Tiger bureaucracy along with many conservationists, like Arjan Singh, wanted the area to be a secure and well-protected national park, but the villagers of Nighasan district have sought to use the forest and adjacent land for forest extraction and agriculture. Neither group has prevailed and the area has been devoted exclusively neither to wildlife conservation nor to agriculture and forestry.

By 1980, Project Tiger was declared a success by some conservationists and the GOI. In 1981 Kailash Sankhala stated that "[i]n the process of saving the tiger, the reserves we so fanatically guarded turned into veritable Gardens of Eden."[139] The number of tigers had grown substantially, the population doubling in some reserves, according to official statistics.[140] In Corbett National Park, the tiger population, according to Project Tiger officials, rose from 46 in 1973 to 100 in 1989.[141] The official tiger population statistics, however, have been disputed on two grounds. Firstly, Arjan Singh has argued that the tiger population increase was due, in part, to the migration of tigers from Nepal into northern India.[142] Secondly, conservationists and wildlife biologists have criticized the methods employed in conducting the census by the Project Tiger staff.[143] Furthermore, critics have suggested that the census figures cannot be scientifically correlated with, for example, an minuscule population of prey species, such as deer, available to support the numbers being claimed in tiger census figures.[144] While it is possible that the tiger population did increase from 1972 to 1980, the census figures are not reliable and the exact number of tigers is not known.[145] By the late 1980s, however, the official statistics indicated a sharp decrease in the tiger population.[146] This decrease has had wider implications for conservationists. Since the tiger serves as an indicator species, any decline in the tiger population suggests that the larger ecosystem is under increased pressure from poachers and the encroachments of landgrabbers.[147] Poaching continued to occur as the trade in wild animal products like fur, ivory, musk, tiger fat and bones, feathers, horns, and meat, flourished in the 1970s and 1980s despite the conservation legislation of both India and Nepal.[148] Other animal populations, like elephants, were under pressure for similar reasons.[149] In 1994 the Zoological Survey of India called for stricter enforcement of the wildlife protection laws and for improvement in the separation of humans and animals through enforcement of the core-buffer zones of Project Tiger sites.[150]

With the creation of Project Tiger and the nationalization of wildlife conservation after 1972, the wildlife in the tarai ceased to serve as a marker for the unusual or exotic. Corbett National Park and Dudhwa National Park were two of an eventual 19 Project Tiger sites scattered throughout the country, and they faced the same hazards, such as man-eating and cattle-lifting tigers and inadequate core-buffer zone structures, as the other reserves. Furthermore, the wildlife populations in forested areas outside of the parks in the tarai dropped significantly after 1947, so that it reached levels found elsewhere in northern India.

Forestry and Afforestation of Land in the Tarai

Not all the forest land in the tarai was reclaimed and transformed into farm land. The tarai's reserved and protected forests, established in the nineteenth century, were maintained and operated by the Forest Department. In Nainital district, much of the tarai forests had been reclaimed for agriculture, but timber operations were still conducted by the Forest Department in the district. One notable forest involved in timber operations was that of Corbett

National Park itself.[151] The park's sal forests were commercially valuable, and timber operations were not halted until the passage of the Wildlife (Protection) Act of 1972. According to Gee, with the active timber operations, the park was "not much more than a very fine reserved forest permanently closed to shooting."[152] The GOI took additional steps to protect the forests with the enactment of the Forest Conservation Act in 1980.[153] Despite this legislation, illegal timber cutting or timber poaching continued in the 1970s and 1980s. In Nainital district, Forest Department officials were unable to stop it. Some officials were involved in the illegal timber trade, but over 48 Forest Department officials were killed by the poachers.[154] In Pilibhit and Kheri districts, timber poachers were also active, and some of them enjoyed political patronage.[155] Forest Department officials often avoided entering forests because of fears that the forests were harboring Sikh "militants." Some of the killings of Forest Department officials committed by timber poachers were officially attributed to the "militants".[156]

The North Kheri Forest Division in tahsil Nighasan was a major site for timber activities until 1977 when it became Dudhwa National Park, although some forestry operations have been undertaken by the Forest Department on occasion throughout the 1970s and 1980s.[157] The timber operations involved mostly native species like sal and netted the Forest Department a profit continually from 1916-17 to 1952-53.[158] In the late 1940s and early 1950s a debate emerged between advocates of different preferred landscapes for the tarai with the Forest Department management divided about the fate of the forests in Nighasan tahsil. In 1949, one Ram Sarup wrote a letter to the editor to the *Indian Forester* and charged the Forest Department with the mismanagement of the North Kheri Forest Division. He stated that it was a "standing disgrace" that there were extensive grasslands (phantas) in the division in which past afforestation efforts had failed. He argued that afforestation efforts in the Tarai and Bhabar Forest Division (Nainital district) had been successful and that efforts in the geographically similar North Kheri Forest Division should be successful as well. He then suggested that the Forest Department should initiate afforestation programs of semal and pakhar (a species of *Ficus*).[159]

In 1952, the *Indian Forester* received two letters, one from a current and one from a retired Forest Department official in response to Sarup. Gerald Trevor, a retired silvicultural officer responsible for past efforts to afforest the phantas, disagreed with Sarup. He argued that the physical terrain of the phantas prevented the growth of trees and that the past efforts to afforest them were doomed to failure from the start.[160] In contrast, James Stephens, IFS, Conservator of Forests, Kumaun Circle, UP, argued that Trevor had misinterpreted the physical landscape of the phantas. He stated that with the use of modern equipment, like tractors, the grasslands could indeed be afforested. He noted that because the colonization project in Kichha tahsil, district Nainital, had been successful with the use of modern equipment, afforestation in the North Kheri phantas should also be successful.[161]

One peculiar aspect of Forest Department activities in the tarai was the reservation of privately owned forests in Kheri and Pilibhit districts. This was done as part of the UP Zamindari and Land Reform Act of 1952.[162] In Nighasan tahsil, Kheri district, about 8,722.04 acres were reserved in 1956.[163] In Puranpur tahsil, Pilibhit district, about 10,516.09 acres were reserved in 1955.[164] In the working plan for the newly reserved forests in Pilibhit, the Forest Department argued that the forests had been heavily damaged under the zamindars and that new management strategies were needed.[165] In addition, the Forest Department promoted the establishment of shooting blocks in some of the newly reserved forests.[166] Of course, the Forest Department's imagined landscape of the tarai included hunting because regulation of hunting in reserved forests earned revenue.

After the reclamation of land in the tarai had been completed, the UP government began the afforestation of land which would serve as fuel and fodder reserves for the colonists. The government's efforts focused many on commercial native species such as sal and teak, but also included exotic species like poplar, eucalyptus, and ipil-ipil. It should be noted that the use of exotic species in afforestation was not limited to UP. Most state forest departments experimented with the use of exotic species, such as eucalyptus, that were thought to be commercially profitable. Poplar trees, imported from the United Kingdom and Italy, were first introduced in the temperate areas of the Himalayan foothills in Nainital district, but the Forest Department began to experiment with their introduction in the tarai itself in 1959.[167] The Forest Department initially conducted research with poplars at two research nurseries at the tarai towns of Lalkua and Clutterbuckganj.[168] At first the Forest Department experimented with a handful of trees, but by 1983 there were some 1100 hectares of poplars.[169] Poplar was selected because of its suitability as a commercial timber crop, but because of the limitations on the cultivation of poplars placed by the tarai climate, the planting of poplars was done in small areas bordering irrigation channels. In addition the poplars required a great deal of care, including irrigation, fertilizing, pruning, and pest control. The Forest Department experimented with the planting of different poplar hybrids under a variety of conditions.[170] The Forest Department, however, ran into a number of difficulties, such as termites, diseases, and the sun-scorch of trees, but none of these problems proved to be unmanageable.[171]

The use of another imported species, eucalyptus hybrids, in tarai afforestation programs began in 1962 at sites in Pilibhit and Nainital districts.[172] The Forest Department chose to experiment with eucalyptus because of its suitability as a commercial timber species and its fast growth rate.[173] The experiments showed that while the soil and climate of the tarai were not uniformly suited for eucalyptus silviculture, some eucalyptus plantations should be established.[174] Eucalyptus afforestation projects in the tarai and throughout India, however, generated a great deal of controversy. While foresters supported the use of eucalyptus, environmentalists, small farmers, and forest dependent communities expressed concern over the economic and environmental effects of the species. Opponents of eucalyptus argued that the species harmed village agricultural economies by replacing useful species like sal and neem with a tree that did not provide fruit, nuts, animal fodder, or green fertilizer; and that in arid areas, eucalyptus unduly lowered the water table because of its high rate of water absorption. Furthermore, grass does not grow under eucalyptus trees and, therefore, eucalyptus plantations led to an additional decline in land suitable for animal grazing since common and grazing lands have been enclosed so that many small cultivators have lost access to important resources.[175] In addition, many eucalyptus plantations were located in areas with high rates of rainfall, like the tarai, that were not environmentally suitable with the result that they were often complete failures.[176] In economic and social terms, the social forestry programs promoted by the various state governments tended to benefit farmers with holdings large enough to support monocultural eucalyptus plantations.[177]

Beginning in 1977, the UP Forest Department experimented with the introduction of ipil-ipil (*Leucaena Leucocephala*) imported from Mexico. This species had a reputation as being useful as a source of paper, timber, and animal fodder and had been grown in Indonesia, Malaysia, and the Philippines. The Forest Department's experiments at the nursery at Lalkua in the Nainital tarai indicated that ipil-ipil had potential as a plantation species in India.[178] Afforestation projects were conducted in the tarai of Kheri district as well. By 1959, the Forest Department had established plantations of some 1,807 acres in the area.[179]

Malaria Control and Malaria Eradication

In the 1960s, the incidence of malaria in the tarai, and throughout India, significantly rose. Malaria was a serious public health problem in the tarai in

1947, but, after several years of malaria control efforts, the tarai was considered by the UP government to be safe, that is, relatively free of malaria. Essentially, this represents a return of malaria to an area after a period of successful control in which the incidence of malaria morbidity declined steadily over three or more years. This phenomenon of the return of malaria to an area after several years of control has two parts: on the one hand, the development of resistance to DDT, BHC, and other chemicals by anopheline mosquitoes, and, on the other hand, the development of resistance to paludrine, chloroquine, and other antimalarial drugs by malaria parasites. Resurgent malaria became a problem in India beginning in the early 1960s, although the problem of resistance to DDT in mosquitoes had started in the early 1950s elsewhere in the world.[180]

Immediately after 1947, malaria control in India was conducted on a regional basis, such as in the UP tarai. Malaria control was instituted on a national basis with the initiation of the GOI's National Malaria Control Programme [NMCP] in 1953. The methods of the NMCP were the same as those used in the tarai beginning in 1948: residual spraying of DDT in buildings, paludrine and chloroquine prophylaxis, and use of other insecticides and larvicides. The goal was to break the cycle of malaria transmission where a mosquito would bite a human carrier of malaria, develop the parasites within its digestive system, and then bite another human and pass on the infection. Paludrine and chloroquine prophylaxis was conducted to remove malaria parasites from individuals to prevent them from serving as reservoirs of infection. Of course, paludrine and chloroquine were used to treat those suffering from the ill effects of malaria as well. Various larvicides, like Paris green, were used to reduce the total mosquito population by either killing the mosquito larvae or by preventing the mosquito eggs from hatching. Residual insecticides, like DDT, BHC, dieldrin, and dichlorovos, were used to kill adult female mosquitoes before they could transmit malaria parasites from one human to another.[181] The idea here was not to wipe out the mosquito population, but to significantly reduce the cycle of malaria transmission from human to mosquito to human. This control of malaria was an on-going process that required annual efforts in distributing chloroquine and spraying insecticides and larvicides.

The GOI as well as the World Health Organization, however, became concerned with the phenomenon of DDT resistance in mosquitoes that was arising throughout India and the world in the mid-1950s.[182] There are different types of resistance, such as natural resistance, induced resistance, and behavioral resistance, but the chief issue was induced or acquired resistance.[183] Malariologist G. Livadas defines induced resistance as when "some species of insects, which at the initial phase of the application of a given chemical substance were susceptible to it, [but] after an extensive exposure to the same substance, acquired the ability of resisting its action."[184] In other words, when first used, DDT would kill a large percentage of the mosquitoes which were exposed to it. But after significant levels of exposure, fewer mosquitoes would be killed, and eventually few, if any, would be killed. The only way, then, to kill mosquitoes would be to use a different insecticide.

To achieve lasting results in the reduction of malaria morbidity, malaria experts affiliated with the WHO as well as with the GOI advocated that malaria control programs should be transformed into malaria eradication programs, and the GOI's National Malaria Eradication Programme [NMEP] was established in 1958.[185] Essentially, the NMEP was a more intensive version of the NMCP—the goal was to wipe out malaria by permanently breaking the cycle of human-mosquito-human transmission of malaria parasites in India before DDT resistance in mosquitoes led to a collapse of then current anti-malaria measures. The Malaria Institute of India, in its *Manual of the Eradication Operation*, defines the goals of eradication:

> Malaria Eradication implies the reduction of the parasite reservoir in human populations to such a negligible degree that once it has been achieved, there is no danger of resumption of local transmission. Complete coverage by intensive spraying over three years may be expected to reduce the human beings who harbour parasites to an extremely low number and bring the degree of infection in them to such a low level as to be incapable of inducing transmission....There should be no equivocation about the exact role of a malaria eradication programme. It is quite distinct from a programme aiming at eradication of mosquitoes. After malaria eradication has been attained, mosquitoes will still be there.[186]

The NMEP sought to break the cycle of malaria transmission through the use of chloroquine and primaquine prophylaxis, larvicides, and insecticides, like DDT and BHC. The chief difference between the NMEP and the NMCP was the organizational structure of the programs as well as a sense of haste. The Malaria Institute of India warned that "[i]t should, however, be remembered that the phenomenon of resistance has introduced an element of urgency, in the matter of converting a control programme into one of eradication."[187]

The problem of mosquito resistance to DDT and other insecticides became an issue in the tarai in the mid 1960s. In the mid 1950s, the anti-malaria programs in the UP tarai were merged into the NMCP and then the NMEP, and NMEP units were active in tarai areas from 1958 until the mid 1970s.[188] The NMEP unit based in Rudrapur and the National Institute of Communicable Diseases, Delhi conducted a survey in the Nainital tarai to determine the extent of DDT resistance in mosquitoes which determined that the chief malaria vector, *A. culicifacies*, had developed a measure of resistance to DDT.[189]

The NMEP, however, faced an unanticipated complication with the development of the resistance to chloroquine and other anti-malaria drugs by *Plasmodium falciparum* throughout India in the 1970s. *Plasmodium vivax*, the other principle malaria parasite in India, did not develop resistance to chloroquine.[190] As with DDT resistance in mosquitoes, chloroquine resistance in malaria parasites took the form of a gradual reduction in the sensitivity of the parasite to the drug. The development of resistance was not uniform throughout India so that laboratory tests were required to learn whether the parasites of a given area were sensitive to chloroquine. In areas where *P. falciparum* was resistant to chloroquine, other drugs were used.[191]

In 1979, the Malaria Research Centre conducted a survey in the tarai which discovered that *P. falciparum* was still sensitive to chloroquine. By 1983, however, *P. falciparum* had developed some resistance to chloroquine in parts of northern India, including the tarai, Haryana state and Delhi.[192] Chloroquine, however, was still effective in the majority of cases, but in those cases where it was not effective, other drugs, like quinine, metakelfin, sulfamethopyrazine, and pyrimethamine were used.[193] Most malaria cases in the tarai were due to *P. vivax*, which did not development resistance to chloroquine.[194]

In the 1970s, there was a resurgence in the incidence of malaria morbidity in the tarai (and much of the rest of India as well). Scientists with the Malaria Research Centre argued that this was due to the "failure" of the NMEP.[195] As such, there were three studies conducted in the Nainital tarai in order to determine the extent of malaria morbidity in the region. These studies determined that the rate of malaria morbidity had increased substantially since the success of the malaria control programs in the 1950s and early 1960s.[196] By 1977, the incidence of malaria morbidity had increased throughout India, and the GOI instituted a new program to combat it called the Modified Plan of Operation [MPO].[197]

The MPO represented the abandonment of the concept of malaria eradication and a return to malaria control. The MPO's goals were to eliminate malaria mortality, reduce malaria morbidity, and protect industrial development and the "green revolution" from malaria morbidity and mortality.[198] The MPO continued to use DDT and BHC in areas where the local malaria vector was still susceptible to them. In areas with mosquito resistance, various insecticides like dieldrin, fenitrothion, pirimiphos, and malathion were used.[199] Chloroquine was still used, but to cope with cases of chloroquine resistance in *P. falciparum*, various other drugs like quinine, primaquine, pyrimethamine, sulphalene, sulfphadoxine, pyrimethamine, septran, amodiaquine, and mefloquine were used.[200]

The resurgence of malaria altered both the preferred and physical tarai landscapes. Beginning in the 1950s, the tarai gained a reputation as a healthy and prosperous area. Indeed, GOI malariologists considered malaria to have been eradicated from the tarai by the mid 1960s.[201] The resurgence of malaria caused a great deal of concern in official circles, and the MPO included the objective of protecting "green revolution" sites like the tarai.[202] This concern was not a problem confined to the tarai, but one present throughout much of India.[203] In a way, it was a marker for the loss of the exotic in the tarai and of its integration into India; the tarai was just one more place in India that faced the resurgence of malaria in the 1970s.

By the 1970s, the tarai ceased to be an exotic or wholly distinct region. Formerly, in the 1940s and 1950s, the tarai was treated as an area that required special attention, such as separate malaria eradication units and the Colonization Department's land reclamation efforts. Beginning in the late 1960s, in the areas of malaria control, agricultural development, and wildlife conservation, the tarai was subject to the same government policies and programs as the rest of the country. The NMCP, the NMEP, the MPO, the New Strategy and the HYVP, Project Tiger, and the eucalyptus afforestation effort were nationwide programs that in no way singled out the tarai region as an area requiring a separate policy or approach.

Notes

[1] Malhotra, Shukla and Sharma 1985:57 and Pattanayak, Arora and Sexana 1976:139.
[2] GOUP 1958 The Uttar Pradesh Agricultural University Act, 1958, p. 1.
[3] Randhawa 1980-86, 4:60, 195.
[4] Borlaug 1983:692 and Hazell 1994:173.
[5] Mann 1989:131.
[6] The membership of these three groups frequently overlapped. For example, John W. Mellor at various times was affiliated with the Rockefeller Foundation, a joint Cornell University-USAID project, the International Bank for Reconstruction and Development, the UN's FAO and the United States Department of Agriculture. See Mellor, Weaver, Lele and Simon 1968.
[7] See Acharya 1973:130 for a discussion of "traditional" agriculture in India.
[8] Quoted in Perkins 1980:12. See also Rudra 1978:390.
[9] Harrar, Mangesldorf and Weaver, "Notes on Indian Agriculture" April 11, 1952, p. 6. Rockefeller Archive Center, Rockefeller Foundation, Record Group 6.7, Series II, Box 26, Folder 147.
[10] Harrar, Mangesldorf and Weaver, "Notes on Indian Agriculture" April 11, 1952, pp. 9 and 11. Rockefeller Archive Center, Rockefeller Foundation, Record Group 6.7, Series II, Box 26, Folder 147.

[11] Harrar, Mangesldorf and Weaver, "Notes on Indian Agriculture" April 11, 1952, p. 12. Rockefeller Archive Center, Rockefeller Foundation, Record Group 6.7, Series II, Box 26, Folder 147.
[12] Farmer 1986:180 and Gilpatric 1969,1:6
[13] See Agarwal 1986, Dhanagare 1988, Goldsmith 1988 and Sharma and Poleman 1993.
[14] Whitcombe 1972:64.
[15] Whitcombe 1972:71-72.
[16] Whitcombe 1972:100-109.
[17] Mukherjee 1992:445.
[18] Pray 1984:430.
[19] GOUP 1926 Report on Agriculture in the United Provinces, pp. 66-67.
[20] GOI, Department of Agriculture, S.V. Section, file number 8-1/46-SV, pp. 1-2.
[21] Karve 1961:1081.
[22] ABD file number 145/1951, volume 4, p. 262.
[23] Mukherjee 1992:447.
[24] Singh, Tyagi, Kathuria and Sahni 1971:901-902.
[25] GOI, Institute of Agricultural Research Statistics 1965, volume 13, part 2, pp. lix-lxx.
[26] Lockwood 1972:A113.
[27] Shiva 1993:39.
[28] Agarwal 1994:312, Farmer 1986:176 and Frankel 2005:581.
[29] Borlaug 1983:691.
[30] Perkins 1990:10 and Lele and Goldsmith 1989:306.
[31] Lele and Goldsmith 1989:308.
[32] Lele and Goldsmith 1989:314 and Lockwood, Mukherjee and Shand 1971:2.
[33] Lele and Goldsmith 1989:310.
[34] Rudra 1978:390.
[35] Perkins 1990:13.
[36] Norman E. Borlaug, "A Brief Report on Progress Being Made by the Indian Coordinated Wheat Improvement Program," April 12, 1966, p. 3. Rockefeller Archive Center, Rockefeller Foundation, Record Group 6.7, Series IV, Sub-series 6, Box 84, Folder 545.
[37] Randhawa 1980-86,4:246.
[38] Rudra 1978:390, Sharma 1985:58 and Staples 1992:16-18.
[39] Lockwood, Mukherjee and Shand 1971:1-2.
[40] Lockwood, Mukherjee and Shand 1971:1-4.
[41] Lockwood, Mukherjee and Shand 1971:2.
[42] Lockwood, Mukherjee and Shand 1971:4-5 and Rudra 1978:382.
[43] A.S. Carter, Walter Scott and Clare Porter, "Seed Improvement in India: Yesterday, Today and Tomorrow," 1969, pp. 3-5. Rockefeller Archive Center, Rockefeller Foundation, Record Group 6.7, Series IV, Sub-series 4, Box, 69, Folder 453 and Lockwood, Mukherjee and Shand 1971:7.
[44] Vyas 1968:A13-A14. Vyas also argued that farmers would be unwilling to try the new varieties unless they could depend upon stable grain prices. See also Mouli 1980:36-39.
[45] Cummings and Ray 1969:A7-A9.
[46] Frankel 1969:693-696.
[47] Lockwood, Mukherjee and Shand 1976:vii.
[48] See Ladejinsky 1973, Tyagi 1974, Dasgupta 1977, and Falcon 1970.

[49] This proposal was drawn up by UP state officials as well as H.W. Hannah of the University of Illinois. UP Agricultural University 1963:6. See also Hannah 1957.
[50] GOI, Ministry of Finance 1959:157 and GOI, Ministry of Finance 1961:119. The project for the UP Agricultural University was included in Agreement number 79 and was provided through the University of Illinois. The initial grant was $147,500. See also Hannah 1957.
[51] University of Illinois at Urbana-Champaign, Department of Architecture 1975:3.
[52] University of Illinois at Urbana-Champaign, Department of Architecture 1975:4.
[53] Randhawa 1980-86,4:187.
[54] University of Illinois at Urbana-Champaign, Department of Architecture 1975:4.
[55] Marothia 1976:86 and Suryanarayana 1975:8.
[56] UP Agricultural University 1963:14-19, Suryanarayana 1975:8, Randhawa 1980-86,4:187 and Charanjit Ahuja "One University That Actually Works" *Indian Express*, March 9, 1994, p.3.
[57] UP Agricultural University 1963:27.
[58] University of Illinois at Urbana-Champaign, Department of Architecture 1975:11.
[59] Mouli 1980:36-48, Subbarao 1980:69, Gilpatric 1969,1:32, 80, and Vyas 1968:A13-A14.
[60] "A Special Correspondent" 1973:2127, Marothia 1976:83, and Gilpatric 1969,1:80.
[61] Mehra 1993:5.
[62] Subbarao 1980:69, Mellor 1976:26-27 and Marothia 1976:82-83.
[63] Mukhoti 1966:1210.
[64] Mehra 1993:5.
[65] See Shah and Singh 1969:88. Singh 1973c uses the term "enterprising farmers." From the use of terms like "progressive" and "enterprising" these economists strongly favored the adoption of HYV technology. Gilpatric 1969,1:17 and 32-33 rather arbitrarily defined as "progressive" any farmer who met any three of the following criteria: devoted at least 20 percent of the farm's sown area to HYV seed, irrigated at least 30 percent of the sown area, used chemical fertilizers on at least 20 percent of the sown area, owned a tractor or power thresher and possessed independent irrigation facilities. Most of the "less progressive" farmers were found wanting on the last criterion.
[66] Shah and Singh 1969:89. "Progressive" farmers tended to be better educated than "less progressive" farmers. According to Gilpatric 1969,1:123, 29 percent of "progressive" farmers had high school or college educations while only 5 percent of "less progressive" farmers did.
[67] Gilpatric 1969,1:32. According to Gilpatric, 65 percent of "progressive" farmers in the Nainital tarai had access to electricity while only 19 percent of "less progressive" farmers did.
[68] Shah and Singh 1969:89-91 and Agrawal and Jain 1969:45-49.
[69] Hobbs, Hettel, Singh, Singh, Harrington and Fujisaka 1991:9 and 20 and Mellor 1976:26-27.
[70] Marothia 1976:82.
[71] Marothia 1976:86.
[72] Agrawal and Jain 1969:45.
[73] Subbarao:1980:69 and Mouli 1980:45-50.
[74] Khare 1972:1-3, Nene 1972a:1-3, Nene 1972b:87-88, Hobbs, Hettel, Singh, Singh, Harrington and Fujisaka 1991:21-44 and Venkataramani 1992:129.
[75] Gilpatric 1969,1:xv and 15.
[76] Mouli 1980:43-45 and 83-85, Hobbs, Hettel, Singh, Singh, Harrington and Fujisaka 1991:53 and Singh and Dubey 1985:20-27.
[77] Singh, Bhati and Shukla 1970:222 and Latham 1998:7.
[78] Ladejinsky 1973:A133 and Tyagi 1979:44.

[79] Agarwal and Shah 1970:213.
[80] Bhati, Moorti, Singh and Verma 1972:41-42.
[81] Bhati, Moorti and Singh 1974:106.
[82] Nag and Burman 1974:12.
[83] Hasan 1979:162.
[84] Bhati, Moorti and Singh 1974:101. According to Singh, Sharma and Mishra 1970:214 the lack of diffusion of HYV technology to the tribal areas of the tarai "indicates that economic development does not take place at the same rate throughout the economy of an area."
[85] Hasan 1979:209 and Bose 2002:59.
[86] Hasan 1988:162.
[87] Seth 1988:408.
[88] Seth 1988:409.
[89] Hasan 1988:165-166.
[90] Mountfort 1970:112 and Waller 1971:751.
[91] Raghavan 1969:738 and Bisht 1995:29-30.
[92] Gee 1992:160 and Stracey 1957:578.
[93] Gee 1992:95 and Lamba (n.d.):1. Ten percent of the park was about 52.3 square kilometers.
[94] Mountfort 1981:75.
[95] Soni 1969:704.
[96] Mountfort 1970:110 and Futehally 1970:37.
[97] GOI, Indian Board for Wildlife 1970:ix-x and Mountfort 1981:91.
[98] Mountfort 1981:82. The exact figure was 1,827 ± 10 percent.
[99] Mountfort 1981:82-84.
[100] GOI, Indian Board for Wildlife 1970:3-8.
[101] GOI, Indian Board for Wildlife 1970:78.
[102] GOI, Indian Board for Wildlife 1970:11-27.
[103] GOI, Indian Board for Wildlife 1970:29.
[104] Futehally 1973:234.
[105] Panwar 1982:130.
[106] Bisht 1995:31 and Panwar 1982:132.
[107] Tinker 1974:804-805 and Mountfort 1983:32.
[108] Ferguson and Kohl 1987:407-408 and United States Fish and Wildlife Service 1994:1-6.
[109] Bedi 1984:15 and Panwar 1982:135.
[110] Singh 1989:36.
[111] Panwar 1982:135.
[112] Sankhala 1993:167.
[113] Singh 1993b:204.
[114] Singh 1993a:98.
[115] Singh 1993b:204.
[116] Singh 1982:83 and GOI, Indian Board for Wildlife 1972:107-114.
[117] Singh 1986b:2
[118] Sankhala 1993:122.
[119] Sankhala 1993:122.
[120] Singh 1993a:100.
[121] Singh 1993a:99.
[122] Sankhala 1993:168.
[123] Sankhala 1993:168.

[124] Choudhury and Sinha 1980:3. See Singh 1993a, Singh 1987, Singh 1984, Singh 1982 and Singh 1981 for Singh's descriptions of his reintroduction projects.
[125] Sale and Singh 1987:82.
[126] Sale and Singh 1987:83.
[127] Singh 1993a:89.
[128] Singh 1990:435. Jackson 1983:279 notes that "human beings are not normal tiger prey, but accidents can occur when tigers are surprised and attack as a defensive reaction. A few tigers also turn to man-eating...because chance encounters teach them that humans are easy to catch."
[129] Alexander, Prasad and Jahagirdar 1991:5 and Jackson 1983:278-279.
[130] Sankhala 1993:168 and Ward 1993:125.
[131] Choudhury and Singh 1980:3.
[132] Singh 1993a:103-163 includes a discussion of the known man-eaters.
[133] Shukla 1995:170. Shukla gives the figure of 20 man-eaters.
[134] *Cat News* 1989:14, Luoma 1987:63 and Jackson 1983:279.
[135] Ward 1993:131.
[136] *Cat News* 1989:14.
[137] Shukla 1995:153-154.
[138] Shukla 1995:176.
[139] Sankhala 1981:26.
[140] Mountfort 1981:108 and Panwar 1982:137. The 1979 census found 3012 tigers in India. Jackson 1983:278. The 1984 census counted 4,000 tigers. Panwar 1987:113 and Luoma 1987:63.
[141] Singh 1989:36. In 1992, Project Tiger officials claimed that Corbett National Park had 92 tigers. Derek Brown, "Sashaying Through Indian Paradise on a Pooping Pachyderm" *The Guardian*, March 7, 1992, p. 9.
[142] Singh 1993a:91, 96 and Singh 1973b:179.
[143] Karanth 1987:126 and Ward 1993:127. One census technique was the counting of tigers by the observation of tiger pug marks or foot prints. Karanth argues that most of the census takers were not sufficiently trained to be able to differentiate between the pug marks of different tigers. See Dharmakumarsinhji 1959:2-6 and Karanth 1997:350-351 for discussion of tiger census methodology.
[144] Karanth 1987:122-126. One further explanation for the inflated figures is the need for officers in the Project Tiger bureaucracy to meet the planned quotas. See Sankhala 1993:137.
[145] Karanth 1987:126-129.
[146] Behler 1995:6 and Johnsingh and Joshua 1994:135.
[147] Kumar 1985:24.
[148] Thakur 1990:563-568, Heinen and Leisure 1993:231, Panjwani 1994:26-42, GOI, Zoological Survey of India 1994:132 and Kumar and Wright 1999:248.
[149] Johnsingh and Joshua 1994:135.
[150] GOI, Zoological Survey of India 1994:133.
[151] GOI, Ministry of Transport and Communications, Department of Tourism 1961:30.
[152] Gee 1952:7.
[153] Kumar 1985:24-25.
[154] Rawat 1993:130-131.
[155] Shukla 1995:154, Misra 1993a:1422 and Misra 1993b:2059.
[156] Rawat 1993:132

[157] Singh 1986b:2.
[158] Bhatia 1955:174-175.
[159] Sarup 1949:475-476.
[160] Trevor 1952:104.
[161] Stephens 1952:265-266.
[162] Garg 1959:7.
[163] FD file number 820/1955, volume 1, pp. 9-13.
[164] FD file number 809/1955, p. 14.
[165] Garg 1959:12-18.
[166] Garg 1959:62.
[167] Pande 1973:12 and Sen 1965:162.
[168] Pande 1973:14.
[169] Pande 1973:14 and Singh, Rawat, Misra, Fasih, Prasad and Tyagi 1983:675.
[170] Pande 1973:12-13.
[171] Pande 1973:14, Thakur 1978:217-221 and Singh, Rawat, Misra, Fasih, Prasad and Tyagi 1983:675, 689.
[172] Chaturvedi 1973:210 and Guha, Sharma, Pant, Singh, Jain and Karira 1973:349.
[173] Sen 1965:164.
[174] Yadav and Prakash 1969:834.
[175] Rawat 1993:131-135, Seabrook 1996:28-29, Shiva 1989:79-82, 180, 206-207 and Shiva, Bandyopadhyay and Jayal 1985:331. See also Gadgil and Guha 1993:190.
[176] Gadgil 1989:27.
[177] Shiva, Bandyopadhyay and Jayal 1985:331-332.
[178] Shukla 1982:228.
[179] Arya 1980:102.
[180] Ascher 1958:622-623 and Bruce-Chwatt 1988:50-51.
[181] Dieldrin was introduced in 1949 and Dichlorovos was introduced in the 1950s to compliment DDT and BHC. Krishnamurthy, Kalra and Sharma 1967:23 and Ascher 1955:34.
[182] See WHO 1970 and WHO 1992.
[183] Livadas 1957:23 and Busvine 1958:265-266.
[184] Livadas 1957:23.
[185] GOI, Ministry of Health, Family Planning and Urban Development 1969:16. The US provided a great deal of support to the NMEP. The USTCM provided DDT and BHC powder, spraying equipment, laboratory equipment, and technical assistance. The Rockefeller Foundation also provided technical assistance. Pletsch, Gartrell and Hinman 1960:58, Ray 1961:12 and GOI, Ministry of Health annual report for the year 1958-59, pp. 38-39.
[186] GOI, MII 1958:27.
[187] GOI, MII 1958:29.
[188] GOI, Directorate General of Health Services, Directorate of National Malaria Eradication Programme 1986,1: 94-125.
[189] Wattal, Singh, Kalra and Johri 1967:374.
[190] Dwivedi, Sahu, Pattanayak and Roy 1979:55-56 and Narasimham 1988:26.
[191] Narasimham 1988:25-26 and Choudhury, Ghosh, Devi and Malhotra 1983:69.
[192] Choudhury, Ghosh, Devi and Malhotra 1983:63.
[193] Choudhury, Ghosh, Devi and Malhotra 1983:69. At this time, chloroquine resistance was found in 6.78 percent of cases.
[194] Malhotra, Shukla and Sharma 1985:57.

[195] Choudhury, Ghosh, Devi and Malhotra 1983:63.

[196] Sharma, Choudhury, Ansari, Malhotra, Menon, Razdan and Batra 1983:21, Choudhury, Malhotra, Shukla, Ghosh and Sharma 1983:50 and Malhotra Shukla and Sharma 1985:57. The reliability of the GOI's malaria morbidity statistics, however, is questionable. According to Kumar 1994:9 a study by the Malaria Research Center [MCR] found that the NMEP dramatically undercounted the cases of malaria throughout India. At Kichha, Nainital district, the NMEP reported 63 cases, but the MRC found 1,784.

[197] Pattanayak and Roy 1980:1 and Bhargava 1986:157.

[198] Pattanayak and Roy 1980:1.

[199] Narasimham 1988:24 and 33-34 and Pattanayak and Roy 1980:9.

[200] Narasimham 1988:34, Pattanayak and Roy 1980:9 and Malhotra, Shukla and Sharma 1985:59.

[201] Malhotra, Shukla and Sharma 1985:57.

[202] Malhotra, Shukla and Sharma 1985:57.

[203] Radhakrishna Rao "Malaria May Outpace Science: Researchers." *Times of India*, April 18, 1994, p. 18.

• CHAPTER EIGHT •

Conclusion

This study is an analysis of the development of the tarai region of Uttar Pradesh and Uttaranchal. The physical landscape, as reflected in the common land use pattern of these districts, underwent a dramatic shift after Indian independence. This study has sought to add to the understanding of South Asian environmental history by focusing on the post-colonial period rather than the colonial era, and by integrating various developmental processes relating to a small geographical area, like forestry, agriculture, wildlife conservation, and malaria control, into a single analysis.

In 1947, the tarai was a distinctive and unusual part of the landscape of northern India. It was viewed primarily as a land of swamps, forests, and grasslands, and was known as the home of abundant wildlife, malarious mosquitoes, and elusive aboriginals by Indians and Europeans alike. For Europeans, the tarai was either the site for hunting during the winter when malaria was less prevalent or a dangerous obstacle to be crossed during the journey from the plains to the hill stations, like Nainital and Mussoorie, during the summer. For Indians, the tarai represented lush grazing for cattle during the winter months, but also the threat of tiger attacks on both humans and cattle. The chief residents of the tarai were forest-dependent populations or tribals, who themselves represented an element of exoticism to Europeans, Muslims, and caste Hindus. Indian peasants resident in the plains to the south of the tarai feared Tharu and Buxa women as magicians who wielded the evil eye while Europeans saw them as remnants of an older era who were first to be studied and then coaxed into the modern world.[1]

By 1975, however, the tarai had been transformed into a normal or usual part of the northern Indian landscape. No longer did the it require the special efforts of the state's Pubic Health or Colonization Departments to make it habitable for Punjabi, Bengali, and Kumauni cultivators. Indeed, by the mid 1960s, the UP government lost its dominant role in the transformation of the tarai. Independent settlers, large farmers, wildlife and timber poachers, landgrabbers, and others sought to implement their own land use agendas. This represented a further stage in the normalization of the tarai in that the ability of the UP government to maintain effective control in the area was no greater than that in neighboring western UP districts.

To better understand the land use agendas of these groups and individuals, it is necessary to examine the explicit and implicit texts or representations they produced to articulate their varying concepts of the landscape. According to geographer James Duncan, "[a] landscape...is a culturally produced model of how the environment should look."[2] These texts were generated by groups and individuals to explain and justify land use policies and actions that implement those policies. In other words, these texts explicated those ideas and perceptions that guided people's interactions with the terrain.

The normalization of the tarai also involved the loss of the elements, like the vast wild animal population or slight human population, that had marked it as an exotic or unusual landscape. The wild animal population of the tarai was drastically reduced to the levels found in much of northern India, so that wild animals were found only in wildlife sanctuaries; and malaria was virtually eradicated from the area until the nationwide resurgence of malaria in the late 1960s. The agricultural land use pattern of the tarai mirrored that of neighboring western UP districts as well. With the implementation of the New Strategy or "green revolution" in 1965, big (and some small) farmers throughout western UP adopted new agricultural technologies like HYV seeds and petrochemical inputs. The level of mechanization and electrification in Nainital district rivaled that of Meerut and Muzzaffarnagar districts, although it was significantly lower in Kheri and Pilibhit districts.[3] The adoption of "green revolution" technology varied greatly within the tarai much as it did throughout western UP. Furthermore, the level mechanization and electrification was higher in much of the Punjab and Haryana than almost any district in UP.

An important element of the normalization of the tarai was the elimination of separate administrative systems, like the Tarai and Bhabar Government Estates. When the Tarai Colonization Scheme was concluded, the UP government regularized the district governments of the area. The tarai was then fully integrated into the political and administrative machinery of the state, and, like all of northern India, was included in Project Tiger, the National Malaria Control Programme, the National Malaria Eradication Program, the Modified Plan of Operations, and the New Strategy and High Yielding Varieties Programme. The tarai was thus incorporated into postcolonial India and differed little from neighboring districts.

In analyzing the normalization of the tarai, this study has produced a number of suggestions regarding the general environmental history of India.

Firstly, it has become clear that the "green revolution" in India was not particularly revolutionary because of the long history of agricultural research and innovation in India before 1965. Furthermore, some scholars have overemphasized the role of the Ford and Rockefeller Foundations in initiating the "green revolution." Not only had the GOI and the country's various agricultural universities been active in agricultural research long before 1965, but the small and large farmers throughout India who were central to the adoption of the new agricultural technology were accustomed to innovation. It is clear that any study of the "green revolution" must include an analysis of the process of HYV technology adoption by small and large farmers, because the availability of irrigation facilities and credit is as important as the availability of HYV seeds and petrochemical inputs.

Secondly, it is important to link the issues of agricultural development and wildlife conservation. To be successful, wildlife conservation sites must be separated from agricultural land by buffer zones, as at Corbett National Park. At Dudhwa National Park, the lack of a buffer zone around the park leads to human and wild animal interaction. Farmers who locate their farms next to Dudhwa often enter the park to gather forest produce; they sometimes encroach on the forest to extend their holdings. In turn, this loss of habitat further reduces wildlife populations. Furthermore, the proximity of the park to the farms facilitates the movement of wild animals, like sugarcane tigers, from the park to agricultural land. The result has been wild animal attacks on crops, livestock, and humans. Consequently tiger attacks on humans have reduced the commitment of politicians and villagers to wildlife conservation. This issue, however, is not limited to the tarai or even to India. It is an issue throughout North America, Asia, and Africa and should be taken into account when studying the environmental history of any of these areas.

Thirdly, it is important to realize that many different groups and individuals, inside and outside of government, are active in shaping the landscape. These various groups and individuals have had differing land use agendas and have sought to implement competing preferred landscapes. In the tarai, the normalization of the area was accomplished by the unscripted actions and interactions of the GOI, the UP government, small and large farmers, timber and wildlife poachers, landgrabbers, environmentalists, Indian and international NGOs, the UP Agricultural University, aid agencies of foreign governments like the US, and UN agencies like the WHO and UNICEF. It is important to note that individuals, like G.B. Pant, Jim Corbett, Indira Gandhi, and Arjan Singh, have played an important role in the implementation of these competing preferred landscapes. Politicians like Gandhi and Pant were powerful decision makers, but Corbett and Singh influenced and shaped the attitudes and perceptions of the tarai of many people throughout India and the world. Therefore, it is important to analyze these actions and interactions of groups and individuals when looking at environmental issues in other parts of India, for example, the construction of the Narmada and Tehri dams and the establishment of eucalyptus afforestation programs.

Furthermore, many of the issues addressed in this study are not exclusively Indian in scope. Issues like the "green revolution," malaria control, and wildlife conservation have a truly international dimension. The "green revolution" was a process of technological innovation that affected agricultural development in many countries. The initial HYV research was conducted in Mexico, India, Pakistan, and the Philippines by organizations supported by the Ford and Rockefeller Foundations and the US government.[4] The aspects of the "green revolution" discussed here, e.g. the question of adoption of HYV technology by small and big farmers, the development of an indigenous research capacity, and the environmental sustainability of the new technology, were issues in many countries in Asia and Africa.[5]

The phenomenon of resurgent malaria is also a global issue. Indeed, the development of DDT resistance in anopheline mosquitoes first occurred in southern Europe in the late 1940s and soon arose in most of the world. Furthermore, chloroquine resistance in malaria parasites has also been present throughout Asia and Africa. Public health authorities in India, like the Malaria Institute of India, have been closely involved in international efforts to develop new mosquito control technologies and strategies. In addition, the US government and the WHO have provided financial and technical assistance to Indian malaria control programs.

Wildlife conservation has long been an international concern. Several glamorous endangered species, like the tiger and elephant, are located in many countries, and efforts to protect them have been effective only through international cooperation, like the CITES ban of the ivory trade. Regulation of the market for wild animal products, like ivory and tiger organs, has also been world-wide. As a consequence, international conservation NGOs, like the WWF, have been active in India, and the GOI has joined IUCN and CITES.

Notes

[1] Stewart 1865:149, Nesfield 1885:26 and Crooke 1974,2:60.
[2] Duncan 1989:186.
[3] Lockwood 1972:A113, Mouli 1980:83-85 and Hobbs, Hettel, Singh, Singh, Harrington and Fujisaka 1991:53.

[4] Norman E. Borlaug, "A Brief Report on Progress Being Made by the Indian Coordinated Wheat Improvement Program," April 12, 1966, p. 2. Rockefeller Archive Center, Rockefeller Foundation, Record Group 6.7, Series IV, Sub-series 6, Box 84, Folder 545.
[5] See Lele and Goldsmith 1990.

Bibliography

India Office Library, London
National Archive of India, New Delhi
Rockefeller Archive Center, Sleepy Hollow, New York
Uttar Pradesh State Archive, Lucknow

Acharya, S.S. 1973 "Green Revolution and Farm Employment." *Indian Journal of Agricultural Economics.* 28:130-145.

Agarwal, Anil (ed.) 1986 *Developing India's Wasted Lands, A Briefing Paper.* New Delhi: Centre for Science and Environment.

Agarwal, Bina 1994 *A Field of One's Own: Gender and Land Rights in South Asia.* New Delhi: Cambridge University Press.

———. 1986 "Women, Poverty and Agricultural Growth in India." *Journal of Peasant Studies.* 13(4):165-220.

Agnew, John A. and James S. Duncan 1981 "The Transfer of Ideas into Anglo-American Human Geography." *Progress in Human Geography.* 5:42-57.

Agrawal, R.C. and S.C. Jain 1969 "Trends in Capital Formation in Tarai Agriculture." *Indian Journal of Agricultural Economics.* 24:44-49.

Agrawal, R.C. and S.L. Shah 1970 "The Tribal Agriculture of 'Bhotiyas' in a New Setting of Tarai Plains." *Indian Journal of Agricultural Economics.* 25(3):209-213.

Alexander, K.C., R.R. Prasad and M.P. Jahagirdar 1991 *Tribals, Rehabilitation & Development.* Jaipur: Rawat.

Ali, Salim 1985 *The Fall of a Sparrow.* Delhi: Oxford University Press.

Anon. 1948 *Big Game Hunting in India and the Game Animals of India.* Delhi.

Arya, V.D. 1980 *Trends of Urbanisation in Tarai and Bhabar Region of U.P. Since 1951.* Unpublished Thesis. Nainital: Kumaun University.

Ascher, K.R.S. 1958 "A Common Information Source For All Resistance Problems in Biology Is Urgently Needed in View of the Paucity of Effective Countermeasures to Insecticide-Resistance!" *Indian Journal of Malariology.* 12(4):615-625.

———. 1955 "Insect Resistance to Dieldrin." *Rivista di parassitologia.* 16:31-40

Asia Watch 1991 *Encounter in Pilibhit: Summary Executions of Sikhs in Uttar Pradesh, India.* New York.

Atkinson, Edwin T. 1980 *The Himalayan Gazetteer.* 3 volumes. New Delhi: Cosmo Publications. Originally published as volume 10 of the *Gazetteer of the North-Western Provinces*, 1882.

Bains, Ajit Singh 1992 *State Terrorism and Human Rights.* New Delhi: Association of Indian Progressive Study Groups.

Bakshi, S.R. 1991 *Govind Ballabh Pant; The True Gandhian.* New Delhi: Anmol Publications.

Bandopadhyay, Arun 1997 "Towards an Understanding of the Environmental History of India." *The Calcutta Historical Journal.* 16(2):153-165.

Batten, T.H. 1846 "Rohilkund, Its Terai and Irrigation." *Calcutta Review.* 5:121-148.

Bedi, Ramesh 1984 *Corbett National Park.* Translated by Ramesh Deshpande. Delhi: Clarion Books.

Behler, Deborah A. 1995 "Project Tiger." *Wildlife Conservation.* 98:6.

Bhargava, Y.S. 1986 *Concepts of Malariology.* Bikaner: Aravali Publishers.

Bhatia, C.L. 1955 *Working Plan for the North Kheri Forest Division, 1953-54 to 1962-63.* Parts 1 and 2. Allahabad: Government of UP.

Bhati, J.P., T.V. Moorti and L.R. Singh 1974 "Development of Tribal Agriculture in Uttar Pradesh, With Special Reference to Tharu Tribes of Nainital Tarai." *Indian Journal of Economics.* 23:101-106.

Bhati, J.P., T.V. Moorti, L.R. Singh, K.K. Verma 1972 "Income Saving and Economic Rationale of Investment in Tribal Agriculture of Nainital Tarai: A Comparative Study." *Indian Journal of Agricultural Economics.* 27(4):37-42.

Bisht, R.S. 1995 *National Parks of India.* New Delhi: Ministry of Information and Broadcasting, Publications Division.

Booth, Martin 1986 *Carpet Sahib, A Life of Jim Corbett.* Delhi: Oxford University Press.

Borlaug, Norman 1983 "Contributions of Conventional Plant Breeding to Food Production." *Science.* 219:689-693.

Bose, Tarun 2002 *Land and Politics in Tarai.* New Delhi: Indian Social Institute.

Braudel, Fernand 1961 "Alimentation et catégories de l'historie." *Annales, Économies, Sociétés, Civilisations.* 16(4): 723-728.

Breeden, Stanley 1993 "Born to Be Wet." *International Wildlife.* 23(3):20-23.

Bruce-Chwatt, Leonard Jan 1988 "History of Malaria From Prehistory to Eradication" in Walther H. Wernsdorfer and Ian McGregor (eds.) *Malaria: Principles and Practice of Malariology.* Volume 1. New York: Churchill Livingstone.

Burt, B.C. 1914 "Steam Ploughing Experiments in the Aira Estate, Kheri, United Provinces." *Agricultural Journal of India.* 9:1-6.

Burton, R.G. 1989 *The Book of the Tiger.* Dehra Dun: Natraj Publishers. Originally published 1933.

Busvine, James R. 1958 "Resistance of Insects to Insecticides: A Note on Terminology." *Indian Journal of Malariology.* 12(4):265-267.

Butler, S.H. 1901 *Final Settlement of the Kheri Settlement.* Allahabad: Government of UP.

Buttimer, Anne 1978 "Charism and Context: The Challenge of La Geographie" in David Ley and Marwyn S. Samuels (eds.) *Humanistic Geography; Prospects and Problems.* Chicago: Maaroufa Press.

Calder, Ritchie 1954 *Men Against the Jungle.* London: Allen & Unwin.

Cat News 1989 "Managing Big Cats in the Vicinity of Human Settlements." 11: 13-15.

Chakrabarti, A.K. 1955 "Malaria Control—A Vital Element in a Mass Drive for Food Production." Part 3. *Bulletin of the National Society of India for Malaria and other Mosquito-born Diseases.* 4(2):53-58.

Chakrabarti, A.K. and N.N. Singh 1957 "The Probable Causes of Disappearance of *A. minimus* From the Terai Area of the Nainital District of Uttar Pradesh." *Bulletin of the National Society of India for Malaria and Other Mosquito-borne Diseases.* 5(2):82-85.

Chakrabarti, Ranjan 2007 "Introduction" in Ranjan Chakrabarti (ed.) *Situating Environmental History.* New Delhi: Manohar.

Champion, F.W. 1990 *With a Camera in Tiger-Land.* Dehra Dun: Natraj Publishers. Originally published 1927.

———. 1934 "The United Provinces." *Journal of the Bombay Natural History Society.* 37:s:105-111. Also reprinted in *Indian Forester* (1934) 60:774-781.

———. 1933 *The Jungle in Sunlight and Shadow.* London: Chatto & Windus.

Champion, H.G. 1974 "Modern Developments in Forestry in the Western Circle, Uttar Pradesh." *Indian Forester.* 100(12):707-711.

Chand, Diwan 1961 "Malaria" in Uttar Pradesh *Report on the State of Health of Uttar Pradesh, With Particular Reference to Certain Diseases.* Lucknow: Government of UP.

Chand, Kalyana 1975 *Govind Ballabh Pant.* Allahabad: Sadhana Sadan.

Chaturvedi, A.N. 1973 "Rotation in *Eucalyptus Hybrid* Plantations." *Indian Forester.* 99(4):205-210.

Chaturvedi, M.D. 1969 "Wildlife Management in India" in *Conservation in India.* Morges, Switzerland: International Union for the Conservation of Nature and Natural Resources. IUCN Publications New Series, Supplementary Paper No. 17.

———. 1935 *Working Plan for the Tarai and Bhabar Estates, United Provinces, 1934-35 to 1948-49.* 2 volumes. Allahabad: Government of UP.

Choudhury, D.S., S.K. Ghosh, C. Usha Devi and M.S. Malhotra 1983 "Response of *Plasmodium falciparum* to Chloroquine in Delhi, Sonepat district of Haryana and Terai region of Uttar Pradesh." *Indian Journal of Malariology.* 20:63-70.

Choudhury, D.S., M.S. Malhotra, R.P. Shukla, S.K. Ghosh and V.P. Sharma 1983 "Resurgence of Malaria in Gadarpur PHC, District Nainital, Uttar Pradesh." *Indian Journal of Malariology.* 20(1):49-58.

Choudhury, S.R. and J.P. Sinha 1980 "The Kheri Maneaters." *Indian Forester.* 106:2-26.

Claval, Paul and Jean-Pierre Nardy 1968 *Pour le cinquantenaire de la mort de Paul Vidal de la Blache.* Cahiers de Géographie de Besancon, numero 16. Paris: Les Belles Lettres.

Collinson, R.F. and J.L. Anderson 1984 "Problems, Principles and Policy in the Reintroduction of Large Mammals in Conservation Areas." *Acta Zoological Fennica.* 172:169-170.

Colvin, Eliot 1873 *Report on the Settlement of Pillibheet, N.-W. Provinces.* Allahabad: Government of UP.

Corbett, Jim 1989 *My India.* Delhi: Oxford University Press. Originally published 1952.

———. 1988 *Man-Eaters of Kumaon.* Delhi: Oxford University Press. Originally published 1944.

Cronon, William 1993 "The Uses of Environmental History." *Environmental History Review.* 17(3):1-22.

———. 1992 "A Place for Stories: Nature, History, and Narrative." *Journal of American History.* 78:1347-1376.

———. 1983 *Changes in the Land: Indians, Colonists, and the Ecology of New England.* New York: Hill and Wang.

Crooke, W. 1975 *The North-Western Provinces of India, Their History, Ethnology, and Administration.* Delhi: Cosmo Publications. Originally published in 1897.

———. 1974 *The Tribes and Castes of the North Western India.* 4 volumes. Delhi: Cosmo Publications. Originally published in 1896.

———. 1896 *The Tribes and Castes of the North-Western Provinces and Oudh.* 4 volumes. Calcutta: Government of India.

Crosby, Alfred W. 1995 "The Past and Present of Environmental History." *American Historical Review.* 100(4): 1177-1189.

Cummings, Jr., Ralph W. and S.K. Ray 1969a "The New Agricultural Strategy: Its Contribution to 1967-68 Production." *Economic and Political Weekly.* 4(13):A7-A16.

———. 1969b "1968-69 Foodgrain Production; Relative Contribution of Weather and New Technology." *Economic and Political Weekly.* 4(39):A163-A174.

Damodaran, Vinita 2007 "Tribes in Indian History" in Ranjan Chakrabarti (ed.) *Situating Environmental History.* New Delhi: Manohar.

Das, Banarsi 1992 *Uttar Pradesh District Gazetteers: Naini Tal.* Lucknow.

Dasgupta, Biplap 1977 *Agrarian Change and the New Technology in India.* Geneva: United Nations Research Institute for Social Development.

Dhanagare, D.N. 1988 "The Green Revolution and Social Inequalities in Rural India." *Bulletin of Concerned Asian Scholars.* 20(2):2-13.

Dharmakumarsinghji, K.S. 1959 *A Field Guide to Big Game Census in India.* New Delhi: Indian Board for Wildlife. Leaflet No. 2.

Duncan, James S. 1989 "The Power of Place in Kandy, Sri Lanka: 1780-1980." in John A. Agnew and James S. Duncan (eds.) *The Power of Place, Bringing Together Geographical and Sociological Imaginations.* Boston: Unwin Hyman.

Dwivedi, S.R., H. Sahu, S. Pattanayak and R.G. Roy 1979 "Presumptive Treatment of *P. vivax* Malaria in Uttar Pradesh, Punjab, Haryana, Himachal Pradesh, Jammu and Kashmir." *Indian Journal of Medical Research.* 70:54-56.

The Economist 1991 "One Man's Law." (August 10), pp.23-24.

Eisenberg, John F. and John Seidensticker 1976 "Ungulates in Southern Asia: A Consideration of Biomass Estimates For Selected Habitats." *Biological Conservation.* 10:293-308.

Ensminger, Douglas 1962 "Overcoming the Obstacles to Farm Economic Development in the Less-Developed Countries." *Journal of Farm Economics.* 44:1367-1387.

Falcon, Walter P. 1970 "The Green Revolution: Generations of Problems." *American Journal of Agricultural Economics.* 52:698-710.

Farmer, B.H. 1986 "Perspectives on the 'Green Revolution' in South Asia." *Modern Asian Studies.* 20:175-199.

———. 1974 *Agricultural Colonization in India Since Independence.* London: Oxford University Press.

Ferguson, David A. and Steven G. Kohl 1987 "Developing International Tiger Conservation Programs: US Fish and Wildlife Service Cooperation With India and the USSR" in Ronald L. Tilson and Ulysses S. Seal (eds.) *Tigers of the World; The Biology, Biopolitics, Management, and Conservation of an Endangered Species.* Park Ridge, New Jersey: Noyes Publications.

Frankel, Francine R. 2005 *India's Political Economy 1947-2004: The Gradual Revolution.* New Delhi: Oxford University Press.

———. 1969 "India's New Strategy of Agricultural Development; Political Costs of Agrarian Modernization." *Journal of Asian Studies.* 28(4):693-710.

Futehally, Zafar 1973 "Conservation in India: 1972 in Retrospect." *Biological Conservation.* 5:233-234.

———. 1970 "10th General Assembly and 11th Technical Meeting of the International Union for Conservation of Nature and Natural Resources." *Journal of the Bombay Natural History Society.* 67(1):30-39.

Gadgil, Madhav 1989 "Deforestation: Problems and Prospects." *Wasteland News* (Supplement, Volume 4, Number 4).

Gadgil, Madhav and Ramachandra Guha 1993 *This Fissured Land; An Ecological History of India.* Berkeley: University of California Press.

Garg, I.P. 1959 *Working Plan for the Erstwhile Private Forests of Pilibhit, 1957-58 to 1966-67.* Allahabad: Government of UP.

Gee, E.P. 1992 *The Wild Life of India.* New Delhi: HarperCollins. Originally published 1964.

———. 1962 *Why Preserve Wild Life?* New Delhi: Indian Board for Wildlife. Leaflet No. 1. Originally published 1956.

———. 1952 "The Management of India's Wild Life Sanctuaries and National Parks." *Journal of the Bombay Natural History Society.* 51(1):1-18.

Gilpatric, C. (ed.) 1969 *Changing Agriculture and Rural Life in a Region of Northern India; A Study of Progressive Farmers in Northwestern Uttar Pradesh During 1967/68.* 2 volumes. Pantnagar, U.P.: U.P. Agricultural University.

•Bibliography•

Goldsmith, Arthur A. 1988 "Policy Dialogue, Conditionality, and Agricultural Development: Implications of India's Green Revolution." *The Journal of Developing Areas*. 22:179-198.

GOI, Central Tractor Organization 1953 *Central Tractor Organization Operational Statistics, 1949-50 to 1951-52*. New Delhi.

―――, Directorate General of Health Services, Directorate of National Malaria Eradication Programme 1986 *Malaria and it Control in India*. 3 volumes. Delhi.

―――, Indian Board For Wildlife 1972 *Project Tiger; A Planning Proposal for Preservation of Tiger (Panthera tigris tigris Linn.) in India*. New Delhi: Ministry of Agriculture.

―――, Indian Board For Wildlife 1970 *Report of the Expert Committee: Wildlife Conservation in India*. New Delhi: Department of Agriculture, Ministry of Food, Agriculture, Community Development and Co-operation.

―――, Information Services 1950 *Land Reclamation*. Washington, D.C.

―――, Institute of Agricultural Research Statistics 1965 *National Index of Agricultural Field Experiments. Volume 13, part 2: Uttar Pradesh, 1954-59*. New Delhi: Indian Council of Agricultural Research.

―――, Malaria Institute of India 1958 *Manual of the Malaria Eradication Operation*. Faridabad.

―――, Ministry of Finance, Department of Economic Affairs 1961 *Second Report on the Indo-U.S. Technical Co-operation Programme*. Faridabad.

―――, Ministry of Finance, Department of Economic Affairs 1959 *Report on the Indo-US Technical Co-operation Programme*. New Delhi.

―――, Ministry of Health, Family Planning & Urban Development 1969 *Report of the Special Committee to Review the Working of the National Malaria Eradication Programme and to Recommend Measures for Improvement*. New Delhi.

―――, Ministry of Health, Central Health Education Bureau 1961 *Malaria Eradication in India*. New Delhi.

―――, Ministry of Information and Broadcasting 1957 *Land Reclamation*. New Delhi.

―――, Ministry of Information and Broadcasting, Directorate of Advertising and Visual Publicity (n.d.) *Story of C.T.O.* New Delhi.

―――, Ministry of Transport and Communications, Department of Tourism 1961 *Wild Life Sanctuaries in India*. New Delhi.

―――, National Malaria Eradication Programme. Annual Report for the year 1963-64.

GOI, Planning Commission 1957 *Review of the First Five Year Plan.* Delhi

———, Reserve Bank of India 1949 *Co-operative Farming.* Bombay.

———, Wastelands Survey and Reclamation Committee 1962 *Report on Location and Utilization of Wastelands in India. Part X—Uttar Pradesh.* Delhi: Ministry of Food and Agriculture.

———, Zoological Survey of India 1994 *The Red Data Book on Indian Animals. Part 1: Vertebrata (Mammalia, Aves, Reptilia and Amphibia).* Calcutta: Zoological Survey of India.

Government of Madras 1959 *Report on Co-operative Farming, Parts I & II Cooperative Farming in Bombay and Uttar Pradesh States.* Madras.

GOUP 1958 *The Uttar Pradesh Agricultural University Act, 1958.* Lucknow.

———. 1957 *Progress Review of the Uttar Pradesh First Five Year Plan 1951-56.* Lucknow.

———. 1953 *Uttar Pradesh Five Year Plan, A Progress Review.* Lucknow.

———. 1950 *Report on the Progress of Development Programmes in the Uttar Pradesh, 1950.* Allahabad.

———. 1926 *Report on Agriculture in the United Provinces.* (Prepared under the orders of the Government of the United Provinces for the Royal Commission on Indian Agriculture.) Nainital, U.P.

———. Agricultural University 1963 *A New System of Education in India.* Pantnagar.

———, Bureau of Agriculture Information 1953 *Conquest of Tarai; A Pictorial Survey of the Reclamation and Colonisation of the Tarai Tract in the District of Naini Tal.* Lucknow.

———, Bureau of Agriculture Information 1950 *The Plan for Agricultural and Rural Development.* Lucknow.

———, Forest Department 1961 *The Centenary of Forest Administration in Uttar Pradesh, 1861-1961; Souvenir.* Dehra Dun, U.P.

———, Information Directorate (n.d.) *Kheri.* Planning Series—19. Lucknow.

———, Medical and Public Health Department. Annual reports for the years 1952-53, 1953-54.

———, Planning Department 1962 *Second Five Year Plan, Progress Review 1956-61.* Lucknow.

———, Planning Department 1960 *Annual Development Plan, 1960-61.* Lucknow.

GOUP, Planning Department 1959a *Annual Development Plan 1959-60 & Progress of Work in the First Three Years 1956-59.* Lucknow.

———, Planning Department 1959b *Progress Report for 1958-59.* Lucknow.

———, Registrar of Cooperative Societies 1954 *Report on the Working of Cooperative Societies in Uttar Pradesh for the Year 1950-51.* Allahabad.

———, Tarai and Bhabar Development Committee 1947 *Report of the Tarai and Bhabar Development Committee, Appointed by the United Provinces for the Investigation of Land Development and Colonization Schemes for Ex-servicemen.* Allahabad.

Guha, Ramachandra 1989 *The Unquiet Woods; Ecological Change and Peasant Resistance in the Himalaya.* Berkeley: University of California Press. Oxford: Oxford University Press.

Guha, S.R.D., Y.K. Sharma, Rajesh Pant, S.P. Singh, D.K. Jain and B.G. Karira 1973 "Pulping Qualities of *Eucalyptus* 'Hybrid' Grown in U.P." *Indian Forester.* 99(6):349-351.

Hamilton, Walter 1828 *The East-India Gazetteer; Containing Particular Descriptions of the Empires, Kingdoms, Principalities, Provinces, Cities, Towns, Districts, Fortresses, Harbours, Rivers, Lakes, etc. of Hindostan, and the Adjacent Countries, India Beyond the Ganges and the Eastern Archipelago; Together With Sketches of the Manners, Customs, Institutions, Agriculture, Commerce, Manufactures, Revenues, Populations, Castes, Religion, History, etc. of Their Various Inhabitants.* 2 volumes. Second Edition. London: Parbury, Allen, and Co.

———. 1820 *A Geographical, Statistical, and Historical Description of Hindostan, and the Adjacent Countries.* 2 volumes. London: John Murray.

Hannah, Harold W. 1957 *The State Farm, Tarai: Its Progress and Its Possibilities As a Rural University.* University of Illinois College of Agriculture.

Hanson, Haldore, Norman E. Borlaug, and R. Glenn Anderson 1982 *Wheat in the Third World.* Boulder, Colorado: Westview Press.

Harrison, Gordon 1978 *Mosquitoes, Malaria and Man: A History of the Hostilities Since 1880.* New York: E.P. Dutton.

Hasan, Amir 1992 *Tribal Development in India.* Lucknow: Print House (India).

———. 1988 *Tribal Administration in India.* Delhi: B.R. Publishing Corporation.

———. 1979 *The Buxas of the Tarai.* Delhi: B.R. Publishing Co.

———. 1976 "The Problem of Land Alienation Among the Buxas of Nainital." *Kurukshetra, India's Journal of Rural Development.* 25(2):10-11.

Hazell, Peter B.R. 1994 "Rice in India." *National Geographic Research & Exploration.* 19(2):173-183.

Heinen, Joel T. and Blair Leisure 1993 "A New Look at the Himalayan Fur Trade." *Oryx.* 27(4):231-238.

Henderson, John M. 1949 "Comments on Man-Made Malaria in India." *Indian Journal of Malariology.* 3:253-259.

Hobbs, P.R., G.P. Hettel, R.P. Singh, Y. Singh, L. Harrington and S. Fujisaka (eds.) 1991 *Rice-Wheat Cropping Systems in the Tarai Areas of Nainital, Rampur, and Pilibhit Districts in Uttar Pradesh, India.* New Delhi: Indian Council of Agricultural Research/Pantnagar, Nainital: G.B. Pant University of Agriculture and Technology.

Hunter, W.W. (ed.) 1885-87 *The Imperial Gazetteer of India.* 14 Volumes. Second Edition. London: Trubner & Co.

Indian Farm Mechanization 1952 "Govt Loans for Tractors." 3(1):5.

Indian Forester 1933 "Editorial Notes—Game Protection Notes." 59:243.

The Indian Medical Gazette 1951 "Malaria on the Retreat in Terai and Bhabar, U.P. (India)." 86:26-27.

Issaris, P.C., S.N. Rastogi and V. Ramakrishna 1953 "Malaria Transmission in the Tarai, Naini Tal District, Uttar Pradesh, India." *Bulletin World Health Organization.* 9:311-333.

Jackson, Peter 1983 "The Tiger Lives On." *Ambio.* 12(3):278-279.

Jalal, R.S., L.S. Bisht, P.P. Elhance 1984 "Cooperative Farming: Some Critical Relations (A Case Study of Tarai Area of District Nainital U.P.)." *Indian Cooperative Review.* 21(8): 225-234.

Johnsingh, A.J.T. and Justus Joshua 1994 "Conserving Rajaji and Corbett National Parks—The Elephant as a Flagship Species." *Oryx.* 28(2):135-140.

Jones, William 1855 *Report on the Canal Irrigation of Rohilcund; Inclusive of Irrigation of Terai Lands, To Which is Added Some Notes on the Direct Management of Roodurpoor and Guddurpoor.* Roorkhee.

Joshi, P.C. 1969 "Terai Landgrabbers Must Be Curbed." *New Age.* 17:6.

Kala, G.R. 1974 *Memoirs of the Raj.* New Delhi: Mukul Prakashan.

Karanth, K. Ullas 1999 "Counting Tigers, With Confidence" in John Seidensticker, Sarah Christie and Peter Jackson (eds.) *Riding the Tiger: Tiger Conservation in Human-Dominated Landscapes.* Cambridge University Press.

———. 1987 "Tigers in India: A Critical Review of Field Censuses" in Ronald L. Tilson and Ulysses S. Seal (eds.) *Tigers of the World; The Biology, Biopolitics, Management, and Conservation of an Endangered Species.* Park Ridge, New Jersey: Noyes Publications.

Karve, D.G. 1961 "Plans of Agricultural Development in India." *Journal of Farm Economics.* 43:1081-1087.

Kashyap, Subhash C. and Mahesh C. Shah 1988 *Govind Ballabh Pant; Parliamentarian, Statesman and Administrator.* New Delhi: National Publishing House.

Kaur, Jitinder 2000 *Terrorism in the Tarai: A Socio-Political Study.* Delhi: Ajanta Books.

Khare, B.P. 1972 *Insect Pests of Stored Grain and Their Control in Uttar Pradesh.* Pantnagar, Uttar Pradesh: G.B. Pant University of Agriculture and Technology Press. G.B. Pant University, College of Agriculture and Research Station Research Bulletin No. 5.

Krishnamurthy, B.S., R.L. Kalra and S.K. Sharma 1967 "Field Evaluation of 'Dichlorovos' for the Control of Mosquitoes." *Bulletin of the Indian Society for Malaria and Other Communicable Diseases.* 4:23-34.

Kumar, Ashok and Belinda Wright 1999 "Combating Tiger Poaching and Illegal Wildlife Trade in India" in John Seidensticker, Sarah Christie and Peter Jackson (eds.) *Riding the Tiger: Tiger Conservation in Human-Dominated Landscapes.* Cambridge University Press.

Kumar, Girish and B.S. Lamba 1985 *Studies on Migratory Birds and Their Feeding Behavior in Corbett National Park.* Calcutta: Zoological Survey of India. Occasional Paper No. 76.

Kumar, S.M. 1985 "Conservation and Corbett National Park." *Yojana.* 29:24-25.

Ladejinsky, Wolf 1973 "How Green Is the Indian Green Revolution?" *Economic and Political Weekly.* 8(52):A133-A144.

Latham, A.J.H. 1998 *Rice: The Primary Commodity.* New York: Routledge.

Lawrence, Henry W. 1982 "Historic Change in Natural Landscapes: The Experimental View." *Environmental Review.* 6(1):14-37.

Leibhardt, Barbara 1988 "Interpretation and Causal Analysis: Theories in Environmental History." *Environmental Review.* 12(1):23-36.

Lele, Uma and Arthur A. Goldsmith 1989 "The Development of National Agricultural Research Capacity: India's Experience With the Rockefeller Foundation and Its Significance For Africa." *Economic Development and Cultural Change.* 37:305-343.

Livadas, G. 1957 "Malaria Vector Resistance to Insecticide." *Rivista di malariologia.* 36:23-38.

Lockwood, Brian 1972 "Patterns of Investment in Farm Machinery and Equipment." *Economic and Political Weekly.* 7(40):A113-A124.

Lockwood, Brian, P.K. Mukherjee and R.T. Shand 1971 *The High Yielding Varieties Programme in India, Part 1.* New Delhi: Government of India, Planning Commission,

Programme Evaluation Organization/Australian National University, Research School of Pacific Studies.

Lockwood, Brian, P.K. Mukherjee and R.T. Shand 1976 *The High Yielding Varieties Programme in India, Part 2 1970-1975.* New Delhi: Government of India, Planning Commission, Programme Evaluation Organization/Australian National University, Research School of Pacific Studies.

Lowenthal, David and Hugh C. Prince 1965 "English Landscape Tastes." *Geographical Review.* 55:186-222.

Luoma, Jon 1987 "The State of the Tiger." *Audubon.* 89(7):61-63.

Mackay, R.D. 1968 *Have You Shot an Indian Tiger?* New Delhi: Lakshmi Book Store.

MacKenzie, John M. 1988 *The Empire of Nature; Hunting, Conservation and British Imperialism.* Manchester: Manchester University Press.

Maheshwari, J.K., K.K. Singh and S. Saha 1981 *The Ethnobotany of the Tharus of Kheri District, Uttar Pradesh.* Lucknow: National Botanical Research Institute.

Malhotra, M.S., R.P. Shukla and V.P. Sharma 1985 "Studies on the Incidence of Malaria in Gadarpur Town of Terai, Distt. Nainital, U.P." *Indian Journal of Malariology.* 22(1):57-60.

Mann, Prem S. 1989 "Green Revolution Revisited: The Adaption of High Yielding Variety Wheat Seeds in India." *Journal of Development Studies.* 26:131-144.

Marothia, D.K. 1976 "Impact of Increase in Input Prices on Production and Profitability of Major Crops in Tarai." *Indian Journal of Agricultural Economics.* 31:81-86.

Mathur, Ravindra Behari 1957 "The Role of Tractors in Artificial Regeneration in Tarai Areas of Uttar Pradesh." *Indian Forester.* 83:101-105.

Mehra, Chandra 1993 *The Levels of Agricultural Productivity in Kumaun Himalaya; An Inter-Regional Analysis.* Unpublished Thesis. Nainital: Kumaun University.

Mellor, John W. 1976 *The New Economics of Growth; A Strategy for India and the Developing World.* Ithaca, N.Y.: Cornell University Press.

Mellor, John W., Thomas F. Weaver, Uma J. Lele and Sheldon R. Simon 1968 *Developing Rural India; Plan and Practice.* Ithaca, N.Y.: Cornell University Press.

Misra, Amaresh 1993a "Mulayam Singh's New Tune." *Economic and Political Weekly.* 28(27/28):1421-1422.

———. 1993b "Land Struggle in Uttar Pradesh." *Economic and Political Weekly.* 28(39):2059.

Misra, A.N. 1965 "Fertilizer Use Improves Crop Yields in Tarai." *Fertilizer News*. March. pp. 7-13, 19.

Moreland, W.H. 1913 *Notes on the Agricultural Conditions and Problems of the United Provinces and of its Districts. (Revised to 1911.)* Allahabad: The United Provinces Press.

Mouli, K. Chandra 1980 *Agricultural Development and Disparities; A Study of Eighteen Western Districts of Uttar Pradesh*. Delhi: University of Delhi, Agricultural Economics Research Centre. Research Study No. 80/5.

Mountfort, Guy 1983 "Project Tiger: A Review." *Oryx*. 17(1):32-33.

———. 1981 *Saving the Tiger*. New York: Viking Press.

———. 1970 "The Bengal Tiger Enters the Red Book." *Animals*. 13(3):110-112.

Mukherjee, A.K. 1982 *Endangered Animals of India*. Calcutta: Zoological Survey of India.

Mukherjee, S.K. 1992 "Progress of Indian Agriculture: 1900-1980." *Indian Journal of History of Science*. 27(4):445-452.

Mukhoti, Bela 1966 "Agrarian Structure in Relation to Farm Investment Decisions and Agricultural Productivity in a Low-Income Country—The Indian Case." *Journal of Farm Economics*. 48:1210-1215.

Mumford, C.A. 1921 *Assessment Report of Pargana Nanakmata and other Villages of the Tarai Sub-Division in the Naini Tal District*. Allahabad: Government of UP.

Nag, N.G. and B.K. Roy Burman (eds.) 1974 *Census of India Volume I Monograph Series Part V-b(iv) Buksa A Scheduled Tribe in Uttar Pradesh*. New Delhi: Ministry of Home Affairs, Office of the Registrar General.

Narasimham, M.V.V.L. 1988 *National Malaria Eradication Programme*. New Delhi: National Institute of Health and Family Welfare. National Health Programme Series 4.

Neelima, Nikunj 1988 *Pandit Govindballabh Pant: Vyakitnv ev Kritidv*. Patna: Bihar Hindi Grant Akadami.

Nene, Y.L. 1972a *A Survey of Viral Diseases of Pulse Crops in Uttar Pradesh*. Pantnagar, U.P.: G.B. Pant University of Agriculture and Technology Press. G.B. Pant University, College of Agriculture and Experiment Station Research Bulletin No. 4.

———. 1972b "Khaira Disease of Rice (*Oryza sativa* L.)." *Indian Journal of Agricultural Sciences*. 42(2):87-95.

Nesfield, John C. 1885 "Tharus and Bogshas of Upper India." *Calcutta Review*. 80:1-46.

Nevill, H.R. 1922 *District Gazetteers of the United Provinces of Agra and Oudh. Volume XXXIV: Nainital.* Lucknow.

———. 1905 *District Gazetteers of the United Provinces of Agra and Oudh. Volume XLII: Kheri.* Allahabad.

Norton, William 1989 *Explorations in the Understanding of Landscape; A Cultural Geography.* New York: Greenwood Press.

Osmaston, A.E. 1927 *A Forest Flora for Kumaun.* Allahabad: United Provinces Government Press.

Pampana, Emilio 1963 *A Textbook of Malaria Eradication.* London: Oxford University Press.

Pande, D.C. 1973 "Initial Performance of *Populus Deltoides* in the Terai of Uttar Pradesh, India." *Indian Forester.* 99:12-18.

Pande, K.N. 1989 *Changing Pattern of Agricultural Landuse in the Tarai & Bhabar Zone of Kumaun.* Unpublished Thesis. Nainital: Kumaun University.

Pande, Y.D. 1961 "Agriculture in Nainital Tarai and Bhabar." *Geographical Review in India.* 23(2):19-39.

Pandey, Shailaja 1982 *Comparative Analysis of Rural Settlement Pattern in the Mountainous, Bhabar, Tarai Regions of Nainital District, Uttar Pradesh.* Unpublished Thesis. Nainital: Kumaun University.

Pandey, Sudhakar 1987 *Govind Ballabh Pant.* New Delhi: Prakashan Vibhag.

Panjwani, Raj 1994 *Courting Wildlife.* New Delhi: World Wide Fund for Nature—India.

Pant, Govind Ballabh 1996 *Selected Works of Govind Ballabh Pant.* Volume 5. Edited by B.R. Nanda. Delhi: Oxford University Press.

———. 1995 *Selected Works of Govind Ballabh Pant.* Volume 4. Edited by B.R. Nanda. Delhi: Oxford University Press.

———. 1994a *Selected Works of Govind Ballabh Pant.* Volume 3. Edited by B.R. Nanda. Delhi: Oxford University Press.

———. 1994b *Selected Works of Govind Ballabh Pant.* Volume 2. Edited by B.R. Nanda. Delhi: Oxford University Press.

———. 1993 *Selected Works of Govind Ballabh Pant.* Volume 1. Edited by B.R. Nanda. Delhi: Oxford University Press.

Panwar, H.S. 1987 "Project Tiger: The Reserves, the Tigers and Their Future." in Ronald L. Tilson and Ulysses S. Seal (eds.) *Tigers of the World; The Biology, Biopolitics, Management, and Conservation of an Endangered Species.* Park Ridge, New Jersey: Noyes Publications.

———. 1982 "Project Tiger." in V.B. Saharia (ed.) *Wildlife in India.* Dehra Dun: Natraj Publishers.

Panwar, H.S. and W.A. Rodgers 1986 "The Re-Introduction of Large Cats into Wildlife Protected Areas." *Indian Forester.* 112:939-944.

Pattanayak, S., D.D. Arora and M.M. Sexana 1976 "Review of NMEP in India." *National Institute of Health Administration and Education Bulletin.* 9(2):133-141.

Pattanayak, S. and R.G. Roy 1980 "Malaria in India and the Modified Plan of Operations for Its Control." *Journal of Communicable Diseases.* 12(1):1-14.

People's Union for Democratic Rights 1989 *Gentlemen Farmers of the Terai, A Report on the Struggle for Land and State Repression in Nainital.* Delhi.

———. 1988 *The Beasts of the Terai; The Report of a Fact-Finding Team on the Incidents of Rape, Looting and Police Torture on the 20th & 21st October, 1988, in Rudrapur Tehsil (Nainital).* Delhi.

Perkins, John H. 1990 "The Rockefeller Foundation and the Green Revolution, 1941-1956." *Agriculture and Human Values.* 7(2):6-18.

Pletsch, Donald J., F.E. Gartrell, E. Harold Hinman 1960 *A Critical Review of the National Malaria Eradication Program of India.* New Delhi: US Technical Cooperation Mission to India, Health Division.

Prakash, Om 1990 *Land Use and Agricultural Development; A Case of Kashipur Tahsil, District Nainital, U.P.* Unpublished Thesis. Nainital: Kumaun University.

Prakash, Om (ed.) 1979 *Uttar Pradesh District Gazetteers: Kheri.* Lucknow.

Pray, Carl E. 1984 "The Impact of Agricultural Research in British India." *The Journal of Economic History.* 44(2):429-440.

Raghavan, S. 1969 "Wild Life Conservation in India." *Indian Forester.* 95:735-740.

Rahman, J., M.V. Singh, M. Pakrasi 1956 "Malaria Control in the Colonization Scheme, Kashipur, District Naini Tal, U.P. (1949-1954)." *Indian Journal of Malariology.* 10(2):155-163."

Randhawa, M.S. 1980-1986 *A History of Agriculture in India.* 4 volumes. New Delhi: Indian Council of Agricultural Research.

Randhawa, M.S. 1980 "Green Revolution in India—Progress, Problems and Prospect." *Proceedings of the Indian National Science Academy.* Part B. 47(1):1-16.

Rao, B. Ananthaswamy 1955 "Acquired Resistance By Insects to Insecticides; Does It Affect the National Malaria Control Plan for India?" *Bulletin of the National Society of India for Malaria and Other Mosquito-borne Diseases.* 3(3):85-92.

Rastogi, Mahendra Kumar 1987 *Pandit Govind Ballabh Pant; Life and Contribution to Indian Politics.* Nainital: Gyanodaya Prakshan.

Rau, M. Chalapathi 1981 *Govind Ballabh Pant; His Life and Times.* New Delhi: Allied Publishers Private Ltd.

Rawat, Ajay S. 1993 *Man and Forests, The Khatta and Gujjar Settlements of Sub-Himalayan Tarai.* New Delhi: Indus Publishing Co.

———. 1992 *Deforestation and Its Impact on the Jammu Gujjars and Tribes in Sub Himalayan Tarai—A Historical Perspective.* New Delhi: Nehru Memorial Museum and Library. Occasional Papers on Perspectives in Indian Development No. 33.

———. 1988 "History and Development of Tarai-Bhabar Region." in K.S. Valdiya (ed.) *Kumaun, Land and People.* Nainital: Gyanodaya Pradashan.

Ray, A.P. 1961 "Review of the National Malaria Eradication Programme." *Bulletin of the National Society of India for Malaria and Other Mosquito-borne Diseases.* 9:10-19.

Redclift, Michael 1995 "In Our Own Image: The Environment and Society as Global Discourse." *Environment and History.* 1(1):111-123.

Ribeiro, Orlando 1968 "En relisant Vidal de la Blache." *Annales de Géographie.* 77(424):641-662.

Richards, John F., Edward S. Haynes and James R. Hagen 1985 "Changes in the Land and Human Productivity in Northern India, 1870-1970." *Agricultural History.* 50(4):523548.

Robb, P.G. 1976 *The Government of India and Reform; Policies Towards Politics and the Constitution, 1916-1921.* Oxford University Press.

Robertson, J.C. 1930 "A Preliminary Report on an Enquiry into Malaria in the Government Estate in the Tarai of the United Provinces." *Records of the Malaria Survey of India.* 1(2):94-113.

Rudra, Ashok 1978 "Organisation of Agriculture for Rural Development: The Indian Case." *Cambridge Journal of Economics.* 2:381-406.

Sahai, R. 1949 *Working Plan for the North Kheri Forest Division United Provinces, 1943-44 to 1952-53.* Parts 1 and 2. Allahabad: Government of UP.

Saksena, Banarsi Prasad (ed.) 1956 *Historical Papers Relating to Kumaun 1809-1842.* Allahabad: Government Central Records Office. Uttar Pradesh State Records Series, Selections from English Records No. 3.

Sale, John B. and Samar Singh 1987 "Reintroduction of Greater Indian Rhinoceros into Dudhwa National Park." *Oryx.* 21(2):81-84.

Sankhala, Kailash 1993 *Return of the Tiger.* New Delhi: Lustre Press.

———. 1981 "Project Tiger." *Sanctuary.* (Bombay) 1:22-35.

Sarup, Ram 1949 "Standing Disgrace to the Forest Department." *Indian Forester.* 75:474-476.

Sauer, Carl O. 1981 *Selected Essays 1963-1975.* Berkeley.

Sauer, Carol O. 1969 *Land and Life: A Selection From the Writings of Carl Ortwin Sauer.* Berkeley.

Savage, Victor R. 1984 *Western Impressions of Nature and Landscape in Southeast Asia.* Singapore University Press.

Seabrook, Jeremy 1996 "Losing the Wood for the Trees." *New Statesman & Society.* (April 12) pp. 28-29.

Sen, N.N. 1965 "Trend of Development of Forestry in Haldwani, Ramnagar and T&B Divisions of Uttar Pradesh During the Last 50 Years." *Indian Forester.* 91:158-169.

Senior-White, R. 1945 "House Spraying With D.D.T. and With Pyrethrum Extract Compared: First Results." *Journal of the Malaria Institute of India.* 6(1):83-93.

Seth, G.R., B.V. Sukhatme and B. Maruti Ram 1958 "A Survey of Agronomic Research Programmes in India." *Indian Journal of Agricultural Research.* 28:409-468.

Seth, J.L. 1988 "Uttar Pradesh" in K.P. Singh (ed.) *Tribal Development in India; Programmes and Implementation.* New Delhi: Uppal Publishing House.

Shah, S.L. 1969 "Income Saving and Investment of Progressive and Less Progressive farmers in North-Western U.P." *Indian Journal of Agricultural Economics.* 24:141-142.

Shah, S.L. and L.R. Singh 1969 "Capital Formation in Agriculture of the Tarai Region of Uttar Pradesh." *Indian Journal of Agricultural Economics.* 24:87-92.

Shah, Shambhu Prasad 1972 *Govind Ballabh Pant, Ev Jivani.* Delhi: Rajkamal Pradashad Praivat Limited.

Sharma, Miriam 1985 "Caste, Class, and Gender: Production and Reproduction in North India." *Journal of Peasant Studies.* 12(4):57-88.

Sharma, Rita and Thomas T. Poleman 1993 *The New Economics of India's Green Revolution; Income and Employment Diffusion in Uttar Pradesh.* Ithaca, N.Y.: Cornell University Press.

Sharma, V.P., D.S. Choudhury, M.A. Ansari, M.S. Malhotra, P.K.B. Menon, R.K. Razdan and C.P. Batra 1983 "Studies on the True Incidence of Malaria in Kharkhoda (District Sonepat, Haryana) and Kichha (District Nainital, U.P.) Primary Health Centres." *Indian Journal of Malariology.* 20:21-34.

Shiva, Vandana, J. Bandyopadhyay and N.D. Jayal 1985 "Afforestation in India: Problems and Strategies." *Ambio.* 14(6):329-333.

Shiva, Vandana 1993 *Monocultures of the Mind; Perspectives on Biodiversity and Biotechnology.* Dehra Dun: Natraj Publishers.

———. 1989 *Staying Alive; Women, Ecology and Development.* London: Zed Books.

Shukla, N.K. 1982 "A Preliminary Note on Strength Properties of *Leucaena leucocephala* From Lalkua, Tarai Bhabar Forest Division (U.P.)." *Indian Forester.* 108:226-229.

Shukla, Rahul 1995 *Killing Grounds; The Saga of Encounters in Wild.* New Delhi: Siddhi Books/Cosmo Publications.

Simmons, I.G. 1993 *Interpreting Nature; Cultural Constructions of the Environment.* New York: Routledge.

Singh, Arjan 1993a *The Legend of the Maneater.* New Delhi: Ravi Dayal Publisher.

———. 1993b "Dudhwa National Park" in Samuel Israel and Toby Sinclair (eds.) *Indian Wildlife.* Second Edition. Houghton Mifflin Co.

———. 1990 "The Man-Eating Phenomenon—An Ecological Crisis" in Trilok Chandra Majupuria (ed.) *Wildlife Wealth of India (Resources & Management).* Bangkok: Tecpress Service.

———. 1986a "Tiger" in R.E. Hawkins (ed.) *Encyclopedia of Indian Natural History.* Delhi: Oxford University Press/Bombay: Bombay Natural History Society.

———. 1984 *Tiger! Tiger!* London: Jonathan Cape.

———. 1981 *Tara, A Tigress.* Edited by John Moorehead. London: Quartet Books.

Singh, Arjan 1973a *Tiger Haven.* Edited by John Moorehead. London: Macmillan.

———. 1973b "Status and Social Behavior of the North Indian Tiger." *The World's Cats.* 1:176-188.

Singh, D., B.N. Tyagi, O.P. Kathuria and M.L. Sahni 1971 "A Survey of Agricultural Experimentation in India." *Indian Journal of Agricultural Sciences.* 41(11):901-913.

Singh, I.J. 1973c "Enterprising Farmers of Tarai." *Eastern Economist.* 60(15):761-63.

Singh, I.J., J.P. Mishra and J.S. Sharma 1970 "Problems of Economic Development in Tribal Agriculture of Tarai." *Indian Journal of Agricultural Economics.* 25(3):214-217.

Singh, K.K., H.S. Bhati and J.K. Maheshwari 1979 "Survey and Biological Activity of Economic Plants Of Kheri Forests, Uttar Pradesh." *Indian Forester.* 105:534-45.

Singh, L.R. 1965 *The Tarai Region of U.P.; A Study in Human Geography.* Allahabad: Ram Marain Lal Beni Prasad.

———. 1961 "The Drainage Difficulties and Flood Problem in the Tarai Region of U.P." *National Geographer.* 4:49-52.

———. 1956 "The Tharus: A Study in Human Ecology." *The National Geographical Journal of India.* 2:153-66.

Singh, L.R. and K.N. Dubey 1985 *Demographic Development in a Developing Economy: A Case Study of Uttar Pradesh.* Allahabad: Govind Ballabh Pant Social Science Institute. Occasional Paper No. 24.

Singh, L.R., J.P. Bhati and V.C. Shukla 1970 "Agricultural Performance of 'Tharus'—A Tribal Community in Tarai Region of Uttar Pradesh." *Indian Journal of Agricultural Economics.* 25:222.

Singh, Pratap, D.S. Rawat, R.M. Misra, Masarrat Fasih, G. Prasad and B.D.S. Tyagi 1983 "Epidemic Defoliation of Poplars and its Control in Tarai Central Forest Division, Uttar Pradesh." *Indian Forester.* 109:675-693.

Singh, Rahul 1989 "Project Tiger." *The Courier* (UNESCO). 42:35-36.

Singh, S.P. 1973c *Centre-State Relations in Agricultural Development.* Delhi: Vikas Publishing House Pvt. Ltd.

Singh, V.B. 1982 "Human Dimension in Wildlife Management in India." *Indian Forester.* 108:449-454.

———. 1978 "The Elephant in U.P. (India)—A Resurvey of its Status After 10 Years." *Journal of the Bombay Natural History Society.* 75(1):71-82.

Sluyter, Andrew 2002 *Colonialism and Landscape: Postcolonial Theory and Its Applications.* Rowman & Littlefield.

Smythies, E.A. 1936 "The Hailey National Park." *Indian Forester.* 62:467-471.

Smythies, Olive 1961 *Ten Thousand Miles on Elephants.* London: Seeley Service & Co. Ltd.

Soni, R.C. 1969 "International Union for Conservation of Nature and Natural Resources." *Indian Forester.* 95:704-705.

"A Special Correspondent" 1975 "P for Pantnagar." *Eastern Economist.* 65(13):ix-xi.

―――. 1973 "Agricultural Extension: Irrelevance of Models and Systems." *Economic and Political Weekly.* 8(48):2127-2128.

Spillett, J. Juan 1967 "General Wild Life Conservation Problems in India." *Journal of the Bombay Natural History Society.* 63(3):616-628.

Srivastava, R.S. 1950 "Malaria Control Measures in the Tarai Area Under the Tarai Colonization Scheme, Kichha, District Naini Tal: September 1947 to December 1948; First Report." *Indian Journal of Malariology.* 4(2):151-165.

―――. 1947 "Final Report on the Field Trials of 'Paludrine' in Selected Hyperendemic Malarious Areas of Naini Tal Tarai in the United Provinces (11th September to 31st December, 1946)." *Indian Journal of Malariology.* 1(3):361-363.

Srivastava, R.S. and A.K. Chakrabarti 1952 "Malaria Control Measures in the Terai Area Under the Terai Colonization Scheme, Kichha, District Naini Tal: 1949-1951." *Indian Journal of Malariology.* 6(4):381-394.

Srivastava, R.S., A.K. Chakrabarti and N.N. Singh 1955 "Study of Chemical and Photodynamic Deterioration of Dichloro-Diphenyl-Trichloroethane (DDT) When Applied on Solid Surfaces." *Indian Journal of Malariology.* 9(1):27-32.

Staples, Eugene S. 1992 *Forty Years: A Learning Curve; The Ford Foundation Programs in India, 1952-1992.* New Delhi: The Ford Foundation.

Stephen, James 1952 "Afforestation of the North Kheri 'Phantas'." *Indian Forester.* 78:265-266.

Stewart, J.L. 1865 "Notes of Observations on the Boksas of the Bijnour District." *Journal of the Asiatic Society of Bengal.* 34:2(3):147-173.

Stracey, P.D. 1960 *Wild Life Management in India.* New Delhi: Indian Board for Wildlife. Leaflet No. 3.

―――. 1957 "The Problem of Preservation of Wildlife in India Today." *Indian Forester.* 83(10):575-581.

Subbarao, K. 1980 "Institutional Credit, Uncertainty and Adoption of HYV Technology: A Comparison of East U.P. With West U.P." *Indian Journal of Agricultural Economics.* 35(1):69-90.

Suryanarayana, K. 1975 "G.B. Pant University of Agriculture: Birth Place of Green Revolution." *Yojana.* 19(6):8.

Swarup, R. 1991 *Agricultural Economy of Himalayan Region With Special Reference to Kumaon.* Nainital: Gyanodaya Prakashan/G.B. Pant Institute of Himalayan Environment & Development.

Tak, P.C. and B.S. Lamba 1984 *Ecology and Ethology of the Spotted-Deer,* Axis axis axis *(Erxleben) (Artiodactyla: Cervidae).* Calcutta: Zoological Survey of India. Occasional Paper No. 43.

———. 1978 "A Review of Census and Monitoring Techniques of Wild Life Populations and Observations on Relative Abundance of Some Mammals in Corbett National Park." in B.K. Tikader (ed.) *Proceedings of the Workshop on Wild Life Ecology.* Calcutta: Zoological Survey of India.

Tarai Farmers' Federation (n.d.) *Land Ceiling in Tarai & Bhabar.* [pamphlet]

Thakur, M.L. 1978 "The Problem of Termite Damage in Poplars in the Bhabar-Tarai Region of Uttar Pradesh." *Indian Journal of Forestry.* 1:217-222.

Thakur, R.K. 1990 "Wildlife Trade" in Trilok Chandra Majupuria (ed.) *Wildlife Wealth of India (Resources & Management).* Bangkok: Tecpress Service.

Thapar, Valmik 1989 *Tigers, The Secret Life.* Emmaus, Pennsylvania: Rodale Press.

Tinker, Jon 1974 "Will India Save the Tiger?" *New Scientist.* 61:802-805.

Trevor, Gerald 1952 "Letter to the Editor." *Indian Forester.* 78(2):104.

Tucker, Richard P. 1988 "The Depletion of India's Forests under British Imperialism: Planters, Foresters, and Peasants in Assam and Kerala." in Donald Worster (ed.) *The Ends of the Earth: Perspectives on Modern Environmental History.* Cambridge University Press.

Tyagi, B.N. 1979 "A Study of the Impact of Green Revolution on the Regional Development of Agriculture in Uttar Pradesh." *Indian Journal of Agricultural Economics.* 29:44-54.

———. 1974 "A Study of the Green Revolution." *Agricultural Situation in India.* 29:205-208.

University of Illinois at Urbana-Champaign, Department of Architecture 1975 *Campus Development Planning Study: G.B. Pant University of Agriculture and Technology.* Urbana: Graduate Division, Department of Architecture, University of Illinois at Urbana-Champaign.

US Fish and Wildlife Service 1994 *Development of the Wildlife Institute of India: Preserving Wildlife Resources of Global Significance.* Washington, D.C.

Uttar Pradesh Agricultural University 1963 *A New System of Education in India.* Pantnagar.

Venkataramani, G. "Pesticides: Harm Far Outweighs Use" in N. Ravi (ed.) *The Hindu Survey of the Environment, 1992.* Madras.

Vidal de la Blache, Paul 1911 "Les genres de vie dans la géographie humaine." *Annales de Géography.* 20:193-212, 289-304.

———. 1903 "La géographie humaine, ses rapports avec la géographie de la vie." *Revue de synthèse historique.* 7:219-240.

Viswanathan, V. 1942 *Final Settlement Report of the Kheri District.* Allahabad: Government of UP.

Vyas, V.S. 1968 "The New Strategy: Lessons of First Three Years." *Economic and Political Weekly.* 3(43):A9-A14.

Waller, Richard 1971 "Last Chance for the Tiger?" *Animals.* 13(16):748-751.

Ward, Geoffrey C. 1993 *Tiger-Wallahs.* New York: HarperCollins.

Wattal, B.L., N.N. Singh, N.L. Kalra and R.C. Jogri 1967 "A Note on DDT Resistance in *Anopheles culicifacies* Giles in Tarai and Bhabar Tract of District Nainital, Uttar Pradesh." *Bulletin of the Indian Society for Malaria and Other Communicable Diseases.* 4(4):373-374.

Whalley, P. 1991 *British Kumaon; The Law of the Extra Regulation Tracts Subordinate to the Government, N.W.P.* Varanasi: Vishwavidyalaya Prakashan. Originally published 1870.

Whitaker, Romulus 1979 "The Crocodilians of Corbett National Park." *Indian Journal of Forestry.* 2:38-40.

Whitcombe, Elizabeth 1972 *Agrarian Conditions in Northern India, Volume 1: The United Provinces Under British Rule, 1860-1900.* Berkeley: University of California Press.

White, Richard 1992 *Land Use, Environment, and Social Change: The Shaping of Island County, Washington.* University of Washington Press.

Whitehead, Neil L. 1998 "Ecological History and Historical Ecology: Diachronic Modeling Versus Historical Explanation" in William Balée (ed.) *Advances in Historical Ecology.* New York: Columbia University Press.

Williams, Michael 1994 "The Relations of Environmental History and Historical Geography." *Journal of Historical Geography.* 29(1):3-21.

World Health Organization 1992 *Vector Resistance to Pesticides; Fifteenth Report of the WHO Expert Committee on Vector Biology and Control.* Geneva. WHO Technical Report Series No. 818.

———. 1970 *Insecticide Resistance and Vector Control; Seventeenth Report of the WHO Expert Committee on Insecticides.* Geneva. WHO Technical Report Series No. 443.

Worster, Donald 1993 *The Wealth of Nature; Environmental History and the Ecological Imagination.* New York: Oxford University Press.

———. 1990b "Seeing Beyond Culture." *Journal of American History.* 76(4):1142-1147.

———. 1990a "Transformation of the Earth: Toward an Agroecological Perspective in History." *Journal of American History.* 76(4):1087-1106.

Yadav, J.S.P. and Jai Prakash 1969 "Soil Suitability for *Eucalyptus* Hybrid (Syn. *E. tereticornis* Or Mysore Gum) Plantations in Tarai and Bhabar Region of Uttar Pradesh." *Indian Forester.* 95:834-840.

Index

A. fluviatilis, 15, 68
A. minimus, 15, 67, 68, 151
adivasi, 30, 65, 100, 104, 120
afforestation, 8, 74, 80, 86, 130, 131, 136, 146
Afzalgarh, 53, 54, 55, 57
Agra Tenancy Act, 47
anopheline mosquito, 14, 24, 25, 67, 132, 146
Asia Watch, 103, 106, 150
Bazpur, 22, 23, 32, 33, 34, 68, 73
BHC, xi, 15, 16, 21, 132, 133, 134, 135, 142
Bhotiyas, 9, 10, 23, 102, 120, 149
Bisht, R.S., 32, 40, 41, 98
Boas, H.J., 30, 65
Braudel, Fernand, 3
Burton, R.G., 34
Buxas, 9, 10, 12, 22, 23, 24, 28, 29, 30, 31, 32, 53, 54, 65, 81, 92, 100, 102, 104, 120, 143, 159
Cassels, W.S., 29
Central Tractor Organization, xi, 7, 16, 45, 58, 64, 66, 68, 72, 78, 79, 85, 86, 87, 88, 155
Champion, F.W., 20, 35
Champion, H.G., 37
Chaturvedi, M.D., 13, 27
chloroquine, 15, 16, 21, 67, 132, 133, 134, 135, 136, 142, 146
CITES, 147
Congress party, 11, 49, 50
Corbett National Park, 2, 8, 12, 13, 14, 38, 88, 107, 122, 123, 124, 127, 128, 129, 140, 145, 150, 160, 171, 173
Corbett, Jim, 9, 19, 20, 33, 36, 56, 91, 102, 146, 150
Cronon, William, 3, 4
Crooke, W., 30
dalit, 57
Damle Committee, 116
Darwin, 50
Das, A.N., 15

Dasgupta, Saibal, 103, 104, 106
Davis, D., 75
DDT, xi, 5, 15, 16, 18, 21, 45, 51, 67, 68, 69, 132, 133, 134, 135, 142, 146, 170, 173
Dhaulpur, 74
dichlorovos, 133
dieldrin, 133, 135
Dudhwa National Park, 2, 8, 12, 13, 14, 105, 107, 122, 124, 125, 126, 127, 128, 129, 145, 167, 168
Duncan, James, 3, 144
East India Company, 8, 23
Emergency, the, 12, 124
eucalyptus trees, 8, 130, 131, 136, 146
famine, 110
Finlay, W.W., 30
Ford Foundation, 11, 113, 114, 117, 171
Forest Conservation Act, 129
G.B. Pant University of Agriculture and Technology, 120, 159, 160, 163, 172
Gadarpur, 16, 81, 152, 162
Gaddis, 9, 10, 28, 93, 96, 100
Gandhi, Indira, 11, 12, 123, 124, 125, 127, 146
Gandhi, Mohandas, 86
Gee, E.P., 14, 87
Giri, V.V., 82
Green Revolution, 2, 11, 17, 50, 84, 103, 104, 107, 108, 109, 110, 111, 113, 117, 118, 119, 135, 136, 144, 145, 146, 149, 153, 154, 155, 161, 162, 165, 166, 168, 171, 172
Grigg, E.E., 31, 41
Grow More Food Campaign, 17, 51, 79, 95, 112
Gujjars, 9, 10, 12, 24, 101, 102, 166
Hailey National Park, 17, 35, 37, 38, 42, 58, 75, 77, 87, 122, 170
Hailey, Malcolm, 38
Hall, W.T., 56, 75
Harpal Singh Sandhu, 64, 71, 98, 116
Harrar, J.G., 110

High Yielding Varieties Programme, 109, 114, 115, 136, 145, 161
High Yielding Variety, 109, 113, 115, 116, 117, 118, 119, 120, 138, 139, 144, 145, 146, 162, 171
Hukam Singh Visen, 76
Hutchinson, F.H., 51, 56
India, Government of, xi, 2, 7, 11, 13, 14, 16, 17, 18, 35, 37, 41, 45, 46, 51, 53, 57, 58, 59, 60, 64, 69, 70, 73, 77, 78, 79, 81, 82, 84, 85, 86, 87, 89, 96, 105, 107, 108, 109, 111, 112, 113, 114, 115, 116, 117, 122, 123, 127, 129, 132, 133, 135, 136, 137, 138, 139, 140, 141, 142, 145, 146, 147, 153, 155, 156, 157, 161, 166
Indian Agricultural Research Institute, 112, 114
Indian Board for Wildlife, 14, 97, 105, 122, 123, 124, 125, 139, 140, 154, 155, 171
Indian Council on Agricultural Research, 112, 113
insecticide, xi, 45, 119, 133
Integrated Tarai Development Project, 121
Intensive Agricultural Districts Program, 114
Issaris, P.C., 68
IUCN, xi, 122, 123, 147, 152
Jafry, Hasan Abid, 37
Jha, A.N., 57, 78, 116
Johnson, J.N.G., 26
Joshi, D.P., 76
Joshi, P.C., 101
jute, 71
Kala, Govind Ram, 33
Kant, Radha, 51, 52, 55, 64, 71, 72, 80, 86, 87, 97, 98, 99
Kashipur, 22, 50, 53, 54, 56, 57, 68, 69, 72, 73, 78, 98, 99, 101, 165
Katju, K.N., 45, 51, 91
Khatima, 23, 26, 29, 32, 111, 121
Kheri, 1, 7, 9, 12, 16, 20, 21, 22, 25, 27, 28, 29, 34, 41, 42, 53, 54, 57, 59, 82, 89, 92, 93, 95, 100, 102, 103, 104, 120, 121, 122, 124, 126, 127, 129, 130, 132, 144, 151, 152, 157, 162, 163, 165, 169, 171, 172
Kichha, 67, 68, 69, 72, 84, 113, 130, 142, 168, 170
Kishanpur Wildlife Sanctuary, 13, 92, 122
Kumaun, ix, 19, 20, 25, 30, 31, 33, 36, 37, 38, 39, 49, 53, 55, 75, 112, 121, 130, 149, 162, 164, 165, 166, 167
Kumaun Development Board, 75
Lambert, G.B., 49
Land Utilization Act, 82
landgrabbing, 92, 100, 101, 102, 121
Lucknow, ix, 12, 45, 107, 121, 149, 152, 153, 157, 158, 159, 162, 163, 165
Mackay, R.D., 34
maize, 95, 113, 117, 119
malaria, 2, 7, 10, 12, 14, 15, 16, 20, 22, 24, 25, 27, 29, 30, 31, 47, 48, 51, 52, 54, 56, 65, 67, 68, 69, 82, 89, 92, 93, 94, 104, 105, 108, 112, 132, 133, 134, 135, 136, 142, 143, 144, 146
Malaria Institute of India, xi, 16, 67, 69, 85, 134, 142, 147, 156, 167
malathion, 69, 136
Malhotra, M.L., 76, 99
Meston, J.S., 25, 40
Modified Plan of Operation, 135, 136, 145, 165
Moreland, W.H., 21
Morris, R.C., 37
Mukerjee, B.K., 71
Nainital, ix, 1, 7, 10, 14, 15, 16, 19, 20, 21, 22, 23, 24, 27, 28, 29, 30, 39, 41, 52, 60, 61, 63, 64, 65, 66, 67, 68, 69, 70, 72, 73, 74, 76, 77, 80, 81, 82, 83, 84, 85, 87, 88, 89, 93, 97, 98, 99, 100, 101, 103, 109, 112, 113, 116, 118, 120, 121, 129, 130, 131, 132, 134, 135, 139, 142, 143, 144, 149, 150, 151, 152, 157, 159, 162, 163, 164, 165, 166, 168, 171, 173
Nanakmata, 23, 163
National Malaria Control Programme, 15, 83, 132, 134, 136, 145
National Malaria Eradication Programme, 2, 16, 18, 134, 135, 136, 142, 155, 156, 163, 165, 166

National Seeds Corporation, 114, 115
Naxalites, 101
Nehru, Jawaharlal, ix, 10, 49, 86, 116
Nepal, 9, 14, 29, 31, 126, 128
Nethersole, Micky, 93
New Delhi, 12, 45, 89, 101, 107, 114, 122, 123, 149, 150, 151, 153, 154, 155, 156, 159, 160, 161, 163, 164, 165, 166, 167, 168, 171
New Strategy, 114, 115, 120, 121, 136, 144, 145, 154, 172
Nighasan, 21, 28, 82, 92, 93, 94, 95, 96, 105, 125, 126, 127, 129, 130
nilgai, 8, 94, 95
North Kheri Forest Division, 12, 21, 27, 34, 36, 92, 94, 101, 104, 107, 122, 124, 126, 129, 150, 167
North Western Provinces, 40, 41, 111, 150, 153
O'Donnell, S.P., 47
Oudh (Avadh), 9, 19, 27, 39, 40, 41, 45, 111, 153, 163
Palia, 28, 93
paludrine, 15, 55, 67, 68, 132, 133
Pandey, Shailaja, 98, 104
Pant, Govind Ballabh, vii, 10, 11, 12, 16, 17, 33, 39, 42, 45, 46, 47, 48, 49, 50, 51, 52, 53, 55, 56, 57, 58, 59, 60, 63, 64, 74, 76, 80, 81, 83, 84, 86, 87, 91, 99, 104, 107, 108, 109, 116, 118, 141, 146, 150, 152, 158, 160, 163, 164, 166, 168, 169, 171
Pantnagar, 11, 12, 108, 116, 117, 118, 120, 155, 157, 159, 160, 163, 170, 172
Panwar, H.S., 124
Paris green, 68, 133
People's Union for Democratic Rights, xi, 101, 106
Phoolbagh, 73
Pilibhit, 1, 7, 9, 13, 16, 20, 21, 22, 27, 28, 34, 39, 41, 42, 54, 59, 60, 82, 89, 92, 93, 95, 102, 103, 105, 106, 112, 120, 129, 130, 131, 144, 150, 155, 159
Plasmodium falciparum, 16, 135, 136, 152
Plasmodium vivax, 135, 154
Plato, 50
poaching, 13, 14, 33, 91, 96, 97, 105, 108, 124, 126, 129
poplar trees, 8, 130
Public Law 480, 117
Puranpur, 22, 28, 95, 130
quinine, 21, 23, 25, 29, 135, 136
refugees, 12, 17, 45, 57, 59, 61, 63, 70, 71, 74, 92, 93, 95, 97, 98, 101, 104
Reserve Bank of India, 69, 74, 87, 156
rhinoceros, 126
rice, 22, 23, 71, 73, 93, 109, 112, 114, 117, 119
Robertson, J.C., 31, 65
Rockefeller Foundation, 11, 109, 113, 114, 117, 137, 138, 142, 145, 146, 147, 161, 165
Rohilkund, 9, 10, 19, 22, 26, 112, 150
Ross, H.G., 30, 41
Rudrapur, 24, 64, 65, 67, 72, 73, 77, 113, 134, 165
Russell, J.W., 51
Sahay, Vishu, 113
sal (*Shorea robusta*), 8, 26, 129, 130, 131
Sankhala, Kailash, 125, 127
Sauer, Carl O., 3
scheduled tribe, 9, 102, 121
Septa Farms, 81
Sethi, D.R., 52
shikar, 13, 32, 33, 34, 35, 38, 93
Shukla, Rahul, 92, 94, 95, 96, 127
Sikhs, 17, 100, 103, 104, 150
Singh, Datar, 82
Singh, Karan, 123
State Tractor Organization, 66, 72, 81, 83
Stracey, P.D., 96
sugarcane, 26, 29, 73, 77, 93, 111, 118, 120, 127, 145
Sultana, 102, 103, 106
Tarai and Bhabar Development Committee, xii, 18, 52, 53, 56, 57, 58, 60, 61, 63, 74, 80, 83, 91, 92, 104, 108, 119, 158
Tarai and Bhabar Government Estates, xii, 10, 16, 21, 22, 23, 24, 25, 26, 28, 29, 30, 31, 32, 33, 34, 40, 41, 45, 46, 47, 48, 49, 50, 51, 53, 54, 55, 56, 57, 58, 59, 66, 86, 89, 92, 97, 98, 111, 144, 152

Tarai Anusuchit Janjati Vikas, 121
Tarai State Farm, 64, 71, 73, 74, 83, 86, 87, 113, 116
Tarai Vikhas Sakakari Sangh, 73
Tharus, 9, 10, 12, 22, 23, 24, 26, 27, 28, 29, 30, 31, 32, 41, 53, 54, 65, 92, 100, 102, 104, 120, 143, 150, 162, 163, 169
tigers, 1, 8, 9, 13, 14, 19, 20, 21, 27, 34, 36, 37, 39, 75, 77, 88, 94, 95, 96, 123, 125, 126, 127, 128, 140, 141, 145
tubewell, 68
Udham Singh Nagar, 1, 7, 9
United Provinces, xii, 9, 19, 22, 39, 41, 42, 45, 49, 111, 137, 151, 152, 157, 158, 162, 163, 164, 167, 170, 173
UP Colonization Department, 5, 55, 57, 58, 59, 63, 64, 65, 66, 68, 69, 70, 72, 73, 74, 77, 80, 81, 83, 87, 99, 136, 143
UP Forest Department, xi, 9, 10, 22, 24, 26, 27, 28, 35, 42, 54, 58, 66, 75, 76, 77, 80, 84, 88, 92, 120, 122, 125, 129, 130, 131, 132, 157, 167
UP Legislative Council, 42, 46, 47
UP Public Health Department, xi, 69, 77, 89, 158
UP Relief and Rehabilitation Department, xii, 93
US Fish and Wildlife Service, 11, 154, 172
US Technical Cooperation Mission, 11, 165
Uttar Pradesh, i, xi, xii, 1, 7, 12, 49, 50, 83, 89, 91, 102, 105, 107, 108, 110, 112, 113, 116, 136, 143, 149, 150, 151, 152, 153, 154, 155, 156, 157, 158, 159, 160, 162, 163, 164, 165, 167, 168, 169, 171, 172, 173
Uttar Pradesh Agricultural University (G.B. Pant University of Agriculture and Technology), 12, 83, 116, 136, 157, 172
Uttaranchal, i, 7, 143
Vidal de la Blache, Paul, 3, 152
Watal, A.P., 56
Weaver, Warren, 110
wheat, 50, 73, 109, 113, 114, 117, 119
Wild Birds and Animals Protection Act, 35, 36
Wildlife (Protection) Act of 1972, 2, 13, 107, 122, 123, 129
wildlife photography, 35, 36, 37, 38
World Bank, 64, 66, 78, 117
World Health Organization, xii, 7, 15, 34, 64, 68, 69, 77, 83, 133, 142, 146, 147, 159, 173
Worster, Donald, 3, 4, 172
WWF, xii, 11, 123, 124, 147
Wyndham, Percy, 25, 33, 34, 38, 40
zamindar, 39, 71

POSTCOLONIAL STUDIES
Maria C. Zamora, *General Editor*

The recent global reality of both forced and voluntary migrations, massive transfers of population, and traveling and transplanted cultures is seen as part and parcel of the postindustrial, postmodern, postcolonial experience. The Postcolonial Studies series will explore the enormous variety and richness in postcolonial culture and transnational literatures.

The series aims to publish work which explores various facets of the legacy of colonialism including: imperialism, nationalism, representation and resistance, neocolonialism, diaspora, displacement and migratory identities, cultural hybridity, transculturation, translation, exile, geographical and metaphorical borderlands, transnational writing. This series does not define its attentions to any single place, region, or disciplinary approach, and we are interested in books informed by a variety of theoretical perspectives. While seeking the highest standards of scholarship, the Postcolonial Studies series is thus a broad forum for the interrogation of textual, cultural and political postcolonialisms.

The Postcolonial Studies series is committed to interdisciplinary and cross cultural scholarship. The series' scope is primarily in the Humanities and Social Sciences. For example, topics in history, literature, culture, philosophy, religion, visual arts, performing arts, language & linguistics, gender studies, ethnic studies, etc. would be suitable. The series welcomes both individually authored and collaboratively authored books and monographs as well as edited collections of essays. The series will publish manuscripts primarily in English (although secondary references in other languages are certainly acceptable). Page count should be one hundred and twenty pages minimum to two hundred and fifty pages maximum. Proposals from both emerging and established scholars are welcome.

For additional information about this series or for the submission of manuscripts, please contact:

>Maria C. Zamora
>c/o Acquisitions Department
>Peter Lang Publishing
>29 Broadway, 18th floor
>New York, New York 10006

To order other books in this series, please contact our Customer Service Department:
>(800) 770-LANG (within the U.S.)
>(212) 647-7706 (outside the U.S.)
>(212) 647-7707 FAX

Or browse online by series:
>www.peterlang.com